Recruiting Older Youths

Insights from a New Survey of Army Recruits

Bernard D. Rostker, Jacob Alex Klerman, Megan Zander-Cotugno

Prepared for the Office of the Secretary of Defense

The research described in this report was prepared for the Office of the Secretary of Defense (OSD). The research was conducted within the RAND National Defense Research Institute, a federally funded research and development center sponsored by OSD, the Joint Staff, the Unified Combatant Commands, the Navy, the Marine Corps, the defense agencies, and the defense Intelligence Community under Contract W74V8H-06-C-0002.

Library of Congress Cataloging-in-Publication Data is available for this publication.

ISBN: 978-0-8330-8390-6

Cover image: Chief of Staff of the Army, Gen. Raymond T. Odierno, issues the oath of office to a group of future soldiers as basic training recruits from Fort Sam Houston, Texas, prior to the start of the 2012 U.S. Army All American Bowl, in San Antonio, Texas, Jan. 7, 2012 (U.S. Army photo by Sgt. 1st Class John Fries/Released).

RAND OFFICES

SANTA MONICA, CA • WASHINGTON, DC

PITTSBURGH, PA • NEW ORLEANS, LA • JACKSON, MS • BOSTON, MA

CAMBRIDGE, UK • BRUSSELS, BE

www.rand.org

Preface

This report aims to improve understanding of the enlistment decisions of older recruits, those who did not join the Army right after high school—assumedly those older than 20 years of age when they enlisted. Since the advent of the all-volunteer force, much attention has been paid to the behavior of young men and women and the decision process that leads them to decide to enlist or to follow a different path after high school. For most of this period, the Youth Attitude Tracking Survey has provided information about such things as the propensity of young men and women to join the military. Today, the Office of the Under Secretary of Defense for Personnel and Readiness surveys young men and women ages 16 through 21, but little is known about older youths and why they join the military.

There has been very little research on older recruits, who made up 48 percent of recruits across all components and services in 2009. This represents a significant part of all recruits. As a group, they had rejected the idea of serving in the military when they graduated high school but changed their minds and have now decided to join. Surveying 5,000 Army recruits between 2009 and 2010, we found that, as a group, those who decided to join after trying "the world of work" had fared less well in the civilian world than had their general population cohort since leaving high school. We argue that this translated into a perception that they faced poor civilian opportunities. We also found that influential individuals who had earlier recommended against joining had, in a substantial fraction of cases, changed their minds. These findings and the research presented here should be of interest to those charged with recruiting for the Army and for the other military components.

This research was sponsored by the Director of Accession Policy in the Office of the Under Secretary of Defense for Personnel and Readiness and was conducted within the Forces and Resources Policy Center of the RAND National Defense Research Institute, a federally funded research and development center sponsored by the Office of the Secretary of Defense, the Joint Staff, the Unified Combatant Commands, the Navy, the Marine Corps, the defense agencies, and the defense Intelligence Community. For more information on RAND's Forces and Resources Policy Center, see http://www.rand.org/nsrd/ndri/centers/frp.html or contact the director (contact information is provided on the web page).

Contents

Figures

Tables

Summary

Since the advent of the all-volunteer force, little attention has been paid to high school graduates who do not enlist immediately after graduation and do not go to college, e.g., those who seek employment in the private sector of the economy. However, over time, this group has made up a significant and increasing portion of total enlistments. For the Army, this group is very important. Since 2005, the majority of the Army's recruits has not joined directly out of high school but has instead made the decision to join at a later time. Why these recruits initially chose not to join when they had the opportunity after graduating from high school and why they changed their minds several years later and enlisted are the subjects of this study. Given the importance of older recruits to the Army, this report examines what is known about them, their performance during military service, and why they came to join the Army after first choosing another postsecondary path. The results of a survey of 5,000 Army recruits designed to answer this question are presented. Finally, the implications of the survey results are discussed, along with suggestions of ways to gain additional insights by tracking this survey cohort through their Army careers.

Our initial insights into older recruits were gained from administrative records obtained from the Military Entrance Processing Command. These data show that, as a group, older recruits score higher on enlistment qualification tests than the group of recruits that (presumably) join directly out of high school, at ages 16 to 19. The data also show that older recruits have attained higher levels of education. At the time of enlistment, most of the youngest group, those 16 to 19 years of age, was either still in high school or had recently graduated. About one-sixth of all recruits ages 22 to 27 have an associate's degree or higher; the fraction is even higher among the oldest group (ages 28 to 42). As expected, older recruits are more likely to be married than younger recruits.

We also explored the relationship between age at enlistment and military career outcomes. We found that older recruits are slightly more likely to leave military service during basic training than are recruits who join directly out of high school—1 to 1.5 percentage points higher for the oldest group (ages 28–42). But once in service, they are more likely to reenlist than are younger recruits. When promotion is considered, the results are even more favorable for older recruits. Assuming that the service member was still in the service at the time to be considered for promotion, older recruits are several percentage points more likely to be promoted. Taken together, the effects of age at enlistment on retention and promotion suggest that the oldest recruits are much more likely to be promoted to noncommissioned officers after four or six years of service—reaching the rank of E-5. At four years, the combined retention and promotion effect is 6 percentage points higher; at six years, the effect is 4 percentage points. Differences are even larger for slightly younger recruits (25–27)—9 percentage points at four years and 7 percentage points at six years.

The preceding discussion concerns the total effect of age. It simply asks: Are older recruits more likely to be retained? We have already seen that older recruits differ from younger recruits. They have higher enlistment qualification test scores, are more likely to have postsecondary education, and are more likely to have a family. Thus, alternatively, we might ask: Is there a separate net effect of age holding these other observable characteristics constant? Given the other characteristics, should we expect an older recruit to perform better? Our analysis shows that separating out the effect of age of enlistment from other factors dampens the effect on retention and promotion, but the basic pattern remains.

While administrative data yield interesting insights into the performance of older recruits once they join the military, these data do not help us understand why older recruits made the decision to join later than did those who enlisted directly out of high school. Thus, to compare Army recruits who joined soon after leaving high school with those who joined later, RAND developed a new survey instrument that was administered at all five of the Army's basic training bases. In total, we received 5,373 completed surveys, with a greater than 90-percent completion rate for those asked to take the survey.

We gained a number of vital insights from the survey results that could be useful in designing future recruiting programs—particularly if the Army decided to specifically target this group of potential recruits in addition to current efforts directed at the high school and college markets.

Our survey data suggest that the military has become a family business. Eighty-three percent of those surveyed had a close family member who had served in the military. Even more impressive was the fact that almost one-half of our sample had a close family member who had retired from the military, one-third of whom were grandparents and almost one-quarter of whom were uncles. We were particularly interested in the number of recruits who had fathers and mothers serving in the military because a comparable national statistic is available from the Department of Labor's Current Population Survey. Our survey revealed that 38 percent of recruits had fathers and 6 percent had mothers who served in the military—percentages that are many times greater than those for the U.S. population as a whole, in which 8.2 percent and 1.3 percent, respectively, have fathers and mothers serving in the military.

The high school has been a central focus of recruiting since the advent of the all-volunteer force in the 1970s. The respondents to our survey overwhelmingly reported that a recruiter had come to their high school—73 percent. However, the response to this question differed significantly when taking into account the time between high school graduation and enlistment. We divided the time of enlistment into three groups relative to the time the respondent left high school. Significantly more respondents who enlisted soon after graduation reported that a recruiter had been to their high schools.

As previously noted, more than one-half of Army recruits do not join immediately after high school. Some decided to continue their educations. Most graduates in our sample of "late enlistees" indicated that, after graduating high school, they went to college and/or vocational school or to work. Some 38 percent, however, indicated that they just took time off. Of this group that joined later, one-quarter indicated that someone did not want them to enlist, and nearly one-quarter also indicated that they were concerned about the wars in Iraq and Afghanistan. When they did enlist, they indicated that the views of others had become less important to their decision and that they were less concerned about the war, despite the fact that nearly all indicated that they expected to be deployed.

We were particularly interested in the reasons late recruits decided to eventually join. About one-third of those who joined later said there were "no jobs at home," and almost one-half were of the view that the jobs that were available were "dead-end jobs." In addition, older recruits' interactions with the Army differed from those of younger recruits. Older recruits actively sought out Army recruiters. Programs built around the school were much less useful to late recruits. Only 24 percent of older recruits indicated that they made contact through their schools, compared with 73 percent of those who enlisted after high school. Fewer older recruits responded to postings at school—only 16 percent compared to 34 percent of early recruits—and those who connected with the military through job fairs were down from 23 to 12 percent. These older recruits were much more likely to stop by recruiting stations and/or fill out request postcards.

We then weighted our survey results, based on key respondent characteristics, to enable us to compare our group of older recruits with a nationally representative group of American youth who also did not join the Army after high school. In general, we found that youth who ultimately joined the Army had not done as well since leaving high school as the general youth cohort had. They were significantly less likely to attend a two- or four-year college. They were more likely to attend a two-year program, as shown by their postsecondary education graduation rates in the second and third years after high school. However, in the fourth and fifth years after high school, when those attending four-year colleges would receive degrees, the graduation rate for the recruits was substantially below the general youth cohort. There were many more high school dropouts in the Army group and very many more who had enrolled in and passed the General Educational Development examination to receive high school diplomas after their high school classes had graduated. Older recruits also had worked less than the general cohort.

Comparing recruits who joined the military some years after graduating high school and a nationally representative group of American youth suggests that the former are doing less well then they may have expected. These older recruits had tested the world of work and found it wanting. Fewer went to college. A larger number were high school dropouts who later enrolled in and passed the General Educational Development examination to be eligible to join the Army. They consistently worked less than the average youth from the comparison group.

For these young Americans, the Army provided a second chance. For those who joined the regular Army, this was a chance to leave home and start again. They understood that they were likely to be assigned to a combat zone, but this did not dissuade them from seeking out the Army and joining. The question now is: How well did the older recruits we surveyed perform during their terms of service, compared to those who had joined after high school? A follow-up study to see how many completed their first term of service, how many reenlisted, and at what rate they were promoted will answer that question. While this report includes an initial exploration of the performance of older recruits using administrative data, following the recruits we surveyed will allow us to link their performance outcomes to the survey results on attitudes and civilian alternatives. In addition, we will be able to examine how the previous work experience of older recruits affects such measures as attrition rates, reenlistment rates, and promotion rates, possibly providing additional tools to recruiters.

Acknowledgments

The research team was strongly supported from the beginning of this work by the project sponsor, Curt Gilroy, former Director of Procurement Policy in the the Office of the Under Secretary of Defense for Personnel and Readiness.

Grateful thanks are extended to the individuals who shepherded this research and its associated materials through the various required approvals: John Jessup, Juanita Irvin, Dave Henshall, and Caroline Miner from the Office of the Secretary of Defense and Robert O. Simmons from the Defense Manpower Data Center.

LTC Sonya Cable facilitated our research needs at Fort Jackson and provided valuable feedback related to the data collection design. Paul A. Goodspeed from Fort Leonard Wood, SGT Jeffrey Krygowski and SGT Matthew McMurray from Fort Benning, 1SG Jerome Draper from Fort Knox, and Scott Seltzer and SFC James Hutchinson from Fort Sill capably coordinated with us as well.

Thanks go to Jennifer Hawes-Dawson of RAND's Survey Research Group for her guidance on data collection methods and to Jennifer Pevar and Erica Czaja for their dedicated research assistance. The authors want to particularly thank Shanthi Nataraj, who carefully read the final version and made extremely helpful recommendations to ensure clarity in presentation, as well as our formal reviewers, Larry Hanser and Curtis Simon.

The authors gratefully acknowledge the over 5,000 enlisted men and women who voluntarily took time from their critical basic training to provide us with survey data for analysis.

Abbreviations

AA	associate in arts
AFQT	Armed Forces Qualification Test
ASVAB	Armed Services Vocational Aptitude Battery
AVF	All-Volunteer Force
BCT	Basic Combat Training
CPS	Current Population Survey
EHC	event history calendar
FY	fiscal year
GED	General Educational Development
MEPCOM	U.S. Military Entrance Processing Command
MOS	military occupational specialty
NRS FY06	New Recruit Survey, Fiscal Year 2006
NLSY97	National Longitudinal Survey of Youth 1997
OSUT	One-Station Unit Training
POC	point of contact
SSN	Social Security Number
TRADOC	U.S. Army Training and Doctrine Command
USAREC	U.S. Army Recruiting Command

CHAPTER ONE

Introduction

Since the advent of the all-volunteer force, considerable research has focused on the contributions to recruiting of recruiters (Dertouzos, 1985), advertising (Dertouzos, 2009), cash incentive programs (Asch et al., 2010), educational benefits (Dertouzos, 1994), and targeted bonus programs (Warner, Simon, and Payne, 2001). Research has focused on specific markets, such as college-bound high school graduates (Asch, Kilburn, and Klerman, 1999), college students (Kilburn and Asch, 2003), and minorities (Asch, Heaton, and Savych, 2009). Little attention, however, has been paid to the high school graduates who do not enlist and do not go to college. Over time, this group has made up a significant and increasing portion of total enlistments, ranging from about 35 percent in 1992 to about 45 percent in 2009. This report examines why these recruits did not join when they had the opportunity at high school graduation and why they changed their minds and enlisted several years later.

Kilburn and Asch have noted that, "[t]raditionally, the [military] services have targeted the recruitment of those youth who have no immediate plans to attend college" (Kilburn and Asch, 2003, p. xvii). As a result, recruiters place high priority on gaining lists of high school students and access to high school campuses. Attitude tracking surveys, such as the Youth Attitude Tracking Survey and its replacement, the Joint Advertising Market Research and Studies Joint Advertising Tracking System, focus on populations of youth who are in high school and immediately after graduation. Little attention has been paid to the attitudes of older youths, those who have graduated from high school to the world of work and, in some cases, joined the military a number of years after graduation, although this group makes up a large percentage of new recruits.

While high school students are the single largest source of new recruits, large numbers of recruits do not join directly after high school or even in the year or two after high school graduation. Using age 18 as a proxy for high school graduation, Table 1.1 shows the age distribution for recruits during fiscal year (FY) 2009 for all services, by component, based on data from the U.S. Military Entrance Processing Command (MEPCOM). Those above age 20 are assumed to have joined several years after graduating from high school, having spent some time in college or trade school or working. This older group is the least important for the Marine Corps, since 67 percent of its new recruits seem to come directly from high school. For the Army, this group is very important, since the majority of its recruits seem not to join directly out of high school, but rather make the decision to join well after leaving high school.

The pattern in Table 1.1 is not new. Figure 1.1 shows that that the percentage of Army active component recruits ages 17 to 19 has fallen sharply since at least the early 1990s, declining from about 65 percent in FY 1992 to just under 45 percent by FY 2009. The largest increase has been among older recruits, particularly those between 22 and 24 years of age. In recent

Table 1.1
Age at Enlistment (percent)

Component		Age Group					
		16–19	20-42	20–21	22–24	25–27	28–42
All		52	48	21	15	6	6
All active		50	50	23	16	6	5
All reserve[a]		56	44	17	12	6	8
Army	Active	44	56	21	18	8	9
	Reserve[b]	56	44	17	12	6	9
Navy	Active	47	53	25	17	6	4
	Reserve	33	67	22	18	8	18
Air Force	Active	50	50	27	17	6	0
	Reserve[c]	44	56	16	18	13	10
Marine Corps	Active	67	33	19	10	3	1
	Reserve	68	32	18	10	4	1

SOURCE: MEPCOM, FY 2009.

[a] All National Guard and Reserve components.

[b] Army National Guard and Army Reserve.

[c] Air National Guard and Air Force Reserve

Figure 1.1
The Fraction of Army Older Recruits Has Risen Over Time

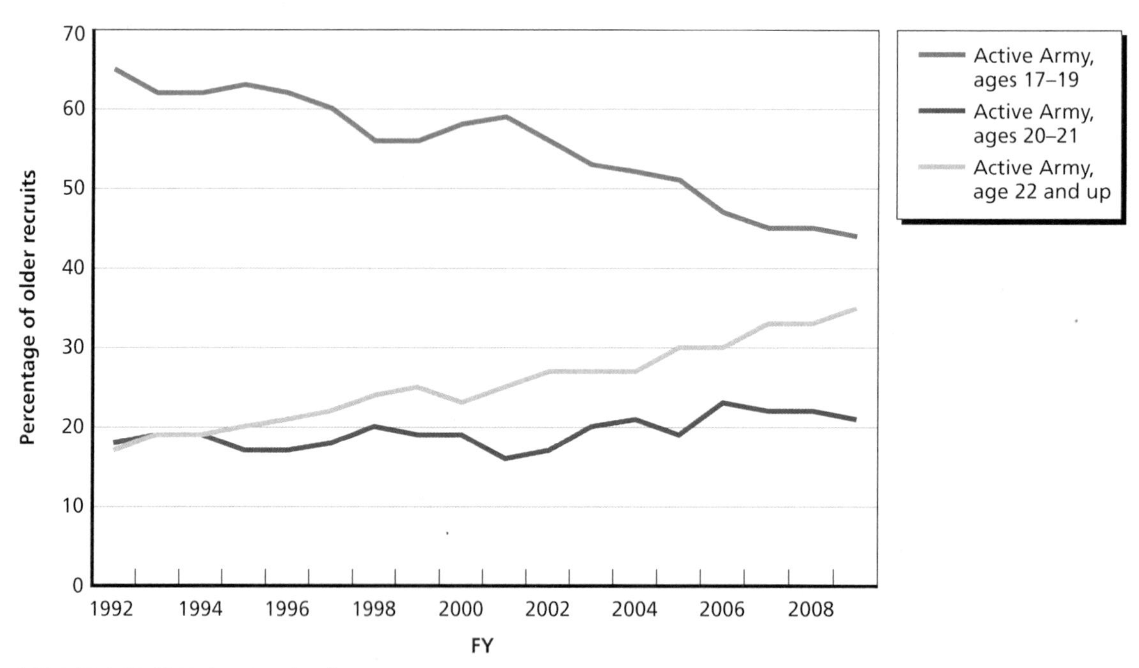

SOURCE: MEPCOM data as of enlistment.

RAND *RR247-1.1*

years, the trend is even more pronounced for the Army Reserve components, as illustrated in Figure 1.2.

Given the importance of older recruits to the Army, the remainder of this report will examine what is known about them, their performance during military service, and why they came to join the Army after first choosing another path after high school. The results of a survey of more than 5,000 Army recruits designed to answer this question are presented, along with a conceptual model of the decision process used to guide the construction of the survey. Finally, the implications of the survey results are discussed, along with suggestions of how tracking the members of this survey cohort through their Army careers might offer additional insights.

Figure 1.2
The Trend Toward Older Recruits Is Found in Both Army Components

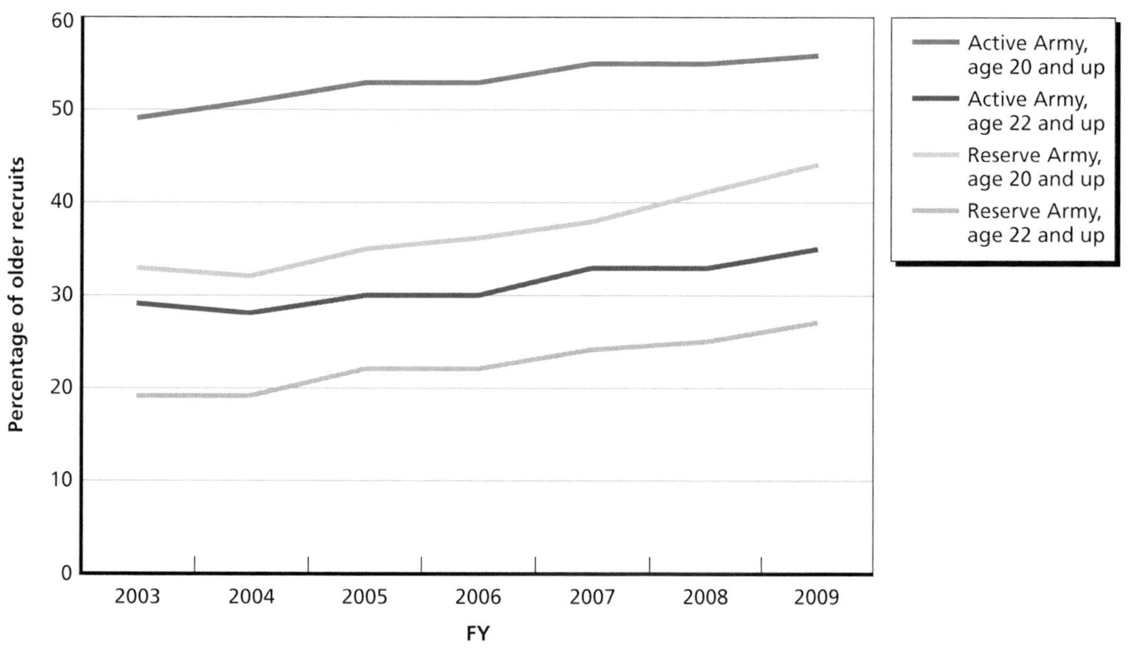

SOURCE: MEPCOM data as of enlistment.
RAND *RR247-1.2*

CHAPTER TWO

Who Are the Older Recruits and How Successful Are They in the Army?

This chapter reviews what administrative records from MEPCOM reveal about older recruits.

Demographics

Table 2.1 presents select statistics for the active Army, using MEPCOM data from FY2009.

Table 2.1
Characteristics of Active Army Recruits, FY 2009 (percent)

	Age Group			
Characteristic	**16–19**	**20–21**	**22–27**	**28–42**
Percentage of total accessions	44	21	26	9
AFQT score				
CAT I–II	39	41	51	50
CAT IIIA	32	26	23	23
CAT IIIB	29	33	26	27
Education				
In high school	29	2	1	1
Alternative high school	15	17	14	12
High school graduate	54	73	61	55
One semester of college	2	5	6	7
Associate's degree or more	0	2	17	25
Race and/or ethnicity				
White	76	77	78	78
Black	14	17	16	16
Hispanic	7	6	6	6
Family structure				
Female	16	14	15	18
Married	5	14	25	56

SOURCE: MEPCOM, FY 2009.

Aptitude Test Results

The Armed Services Vocational Aptitude Battery (ASVAB) is a multiple-choice computer adaptive test used to determine qualification for enlistment in the armed forces. Generally, high school students take it, and high school guidance departments and the military use the results to help guide high school students into career paths for which they have a particular aptitude. The test contains nine sections: Word Knowledge, Arithmetic Reasoning, Mechanical Comprehension, Automotive and Shop Information, Electronics Information, Mathematics Knowledge, General Science, Paragraph Comprehension, and Assembling Objects. Scores on several of these sections are combined to calculate the Armed Forces Qualification Test (AFQT) score, which is divided into percentile categories as follows:

- category I: the 93rd through 99th percentiles
- category II: the 65th through 92nd percentiles
- category IIIA: the 50th through 64th percentiles
- category IIIB: the 31st through 49th percentiles.

In 2004, the test's percentile ranking scoring system was renormalized to ensure that a score of 50 percent represented the median for American youth. The overall goal of the Army is to have at least 60 percent of all recruits score above the 50th percentile, i.e., categories I, II, or IIIA.

The data presented in Table 2.1 show that more older recruits have scores in categories I and II than the recruits assumed to join directly out of high school, i.e., at ages 16 to 19. Accordingly, fewer older recruits have scores in categories IIIA and B.

Education

Most of the youngest group (ages 16 to 19) were either still in high school when they joined or had recently graduated. Very few of them had been to college or attended a technical school. The education attainment for the other age groups shows, as expected, that they are no longer in high school; a few of them have attended college; and a rising fraction has received associates degrees from junior colleges. That older recruits are more likely to have more education should not be surprising. An associate's degree usually requires two years of study. Therefore, even a young person who graduates high school at age 18 and attends a community college full time would not have enough time to receive an associate in arts (AA) degree until age 20. These data suggest that it may take this group somewhat longer than two years to accumulate enough credits to receive an AA degree because many enroll in an AA program as part-time students. Even among 20- and 21-year-olds, AA degrees and even a semester in community college remain rare. However, about one-sixth of older recruits (ages 22 to 27) have AA degrees.

Research starting in the late 1990s suggested to the Army that "the college market" might be a good source of Army recruits (see Asch, Kilburn, and Klerman, 1999, and Kilburn and Asch, 2003), since high quality recruits who were college bound might not successfully complete their college education.[1] As a result, the Army increased its emphasis on high school graduates and encouraged recruiters to spend time on college campuses and with college students. The data reported here do not, however, reflect the effectiveness of this shift in focus, raising a

---

[1] The Department of Defense defines a high-quality recruit as an individual who scores in Category IIIA and above on the AFQT and is a high school graduate.

question about the effectiveness of that approach to recruiting older youths. Most of the older recruits who actually join the Army do not have even a semester of community college credit.

Race or Ethnicity

The ethnic composition of the Army does not change much by entry age cohort.

Family Structure

The MEPCOM data show that older recruits are more likely to be married than younger recruits. This may indicate that older recruits are simply further along in life. There are, however, factors that make the Army attractive to families. Army compensation is slightly higher for a recruit with a family than for a single recruit. For the active recruit, full health care is provided for all family members. Subsidized childcare is also available. In addition, the stable income the military provides may be more attractive for those with family responsibilities. Chapter Four will further explore these and other questions using the survey of recruits.

Experience in the Military

Previous Studies

Age at enlistment has seldom been studied directly, but insights have been gained when entry age has been a controlled variable in several earlier studies of "attrition," i.e., those who fail to complete their initial term of service. Treating age as a continuous variable, Simon, Negrusa and Warner, 2010, found that age had a small effect on attrition after the first two years of military service. They found that older recruits were more likely to complete two years of service, but the effect is only about 0.1 percentage point per year in the Army. Compared to those under age 20, those ages 20 to 23 were 2.4 percent more likely to complete the first term; those age 24 and older were another 2.8 percent more likely to complete the first term. They also found that "probability of separation and education benefit usage were strongly decreasing in age at entry into the military, and were lower for males, married individuals, and individuals with dependents" (Simon, Negrusa, and Warner, 2010, p. 1016).

A New Analysis

Tables 2.2 and 2.3 explore the effects of age at enlistment on military career outcomes using data from the Defense Manpower Data Center's Work Experience File longitudinal data file. These data include both basic demographic information and information about attrition, retention, and promotion, where *attrition* is defined as leaving the force before the contracted term of service; *retention* is defined as continuing past the first term of obligated service; and *promotion* is defined as advancing to a higher grade. We considered nine outcomes:

1. the probability of remaining in service for at least three months, that is, completing basic training
2. retention to four years of service
3. retention to six years of service
4. retention to four years of service conditional on having served at least three months
5. retention to six years of service conditional on having served at least four years

Table 2.2
Total Effects of Age at Entry on Army Retention and Promotion, Relative to 16- to 19-Year-Olds, 1995–2001

	Age at Entry			
	20–21	22–24	25–27	28–42
Retention at 3 months	−0.3	0.2	−0.2	−1.3
Retention at 4 years	−0.1	2.6	3.5	0.9
Retention at 6 years	−0.2	2.8	4.6	2.3
Retention at 4 years for those who served at least 3 months	0.4	2.8	4.2	2.1
Retention at 6 years for those who served at least 4 years	0.0	0.7	1.3	1.8
Promotion to E5 by 4th year for those who served at least 4 years	2.9	8.9	10.9	8.3
Promotion to E5 by 6th year for those who served at least 6 years	1.0	4.1	4.9	2.8
Promotion to E5 by 4 years	1.9	7.3	9.2	6.2
Promotion to E5 by 6 years	0.7	4.8	6.8	3.6

NOTE: Each row of this table contains separate linear probability model estimates of the effects of age at entry on each enlistment outcome. Each number shows the effect, in percentage points, of being in a given age category relative to those ages 16 to 19.

Table 2.3
Total and Partial Effects of Age at Entry on Army Retention and Promotion Relative to 16- to 19-Year-Olds, 1995–2009

	Age at Entry							
	20–21		22–24		25–27		28–42	
	Total	Partial	Total	Partial	Total	Partial	Total	Partial
Retention at 3 months	−0.3	−0.1	0.2	0.3	−0.2	−0.1	−1.3	−0.9
Retention at 4 years	−0.1	0.8	2.6	2.6	3.5	3	0.9	0.8
Retention at 6 years	−0.2	0.4	2.8	2.5	4.6	3.4	2.3	1.7
Retention at 4 years for those who served at least 3 months	0.4	1.1	2.8	2.7	4.2	3.4	2.1	1.5
Retention at 6 years for those who served at least 4 years	0	0	0.7	0.4	1.3	0.6	1.8	1.1
Promotion to E5 by 4th year for those who served at least 4 years	2.9	1.5	8.9	4.3	10.9	4.3	8.3	1.1
Promotion to E5 by 6th year for those who served at least 6 years	1	0.1	4.1	0.2	4.9	−1.1	2.8	−4
Promotion to E5 by 4 years	1.9	1.3	7.3	4	9.2	4.1	6.2	0.9
Promotion to E5 by 6 years	0.7	0.3	4.8	2	6.8	2	3.6	−1.5

NOTE: Each row of this table contains separate linear probability estimates of the effects of age at entry on each enlistment outcome, with the partial effects models controlling for the characteristics contained in Table 2.2. Each number shows the effect, in percentage points, of being in a given age category relative to those ages 16 to 19.

6. the probability of achieving the military grade of E-5 by the fourth year, conditional on remaining in the service at the time to be considered for promotion
7. the probability of achieving the military grade of E-5 by the sixth year, conditional on remaining in the service at the time to be considered for promotion
8. the unconditional probability of achieving the military grade of E-5 at four years of service
9. the unconditional probability of achieving the military grade of E-5 at six years of service.

We examined both the total effects of age and the partial effects of age, holding the characteristics in Table 2.1 constant.

Table 2.2 shows the total effects of age at enlistment on career outcomes for Army enlistees. These effects are computed from separate linear probability regression models that include only dummy variables for each age group (16 to 19 is the excluded category) and (federal fiscal) year of contract. The parameter estimates therefore represent the difference in the probability of retention or promotion, in percentage points, relative to 16- to 19-year-olds. Positive numbers indicate that older recruits are more likely to be retained or promoted than 16-to-19-year-olds, while negative numbers indicate that they are less likely to be retained or promoted than 16-to-19-year-olds.

The results give the cumulative probability of retention at three months, four years, and six years. Early retention during initial basic training is slightly lower for older recruits, 1 to 1.5 percentage points for the oldest group (ages 28 to 42). At the later retention points, however, this pattern reverses, and retention rates are higher for older recruits. When promotion is considered, the results are even more favorable for older recruits. Conditional on staying in the military to four years and six years, older recruits are several percentage points more likely to be promoted. Combining the positive effects of age at enlistment on retention and promotion conditional on retention implies that the oldest group of recruits is much more likely to be promoted to noncommissioned officer (i.e., E-5). For the oldest recruits, the effect at four years is 6 percentage points, and the effect at six years is 4 percentage points. Differences are even larger for slightly younger recruits (ages 25 to 27)—9 percentage points at four years and 7 percentage points at six years.

The preceding discussion concerns the "total effect" of age. It simply asks: Are older recruits more likely to be retained? We have already seen that older recruits differ from younger recruits. They have higher AFQT scores, are more likely to have post–high school education, and are more likely to have a family and dependents. It is therefore important to know whether older recruits are, for example, more interested in housing and medical benefits or whether age proper, holding constant marital and dependent status, is driving the results.

To address this question, we augmented the linear probability models in Table 2.2 to include the characteristics reported in Table 2.1 as independent variables. The resulting estimates, seen in Table 2.3, now yield the *partial* effects of age, holding those other characteristics constant. For ease of comparison, Table 2.3 also reproduces the total effects of age from Table 2.2.[2]

[2] One aspect of the specification is crucial. There are essentially no young recruits with AA degrees (or even one semester of credit). We therefore ran the models both with a complete education specification and, alternatively, with a specification that recodes those with any college (including a degree) as conventional high school graduates. It is the latter specification that is reported here.

Controlling for nonage observable factors dampens the effects of age of enlistment on retention and promotion, but the basic pattern remains. For the Army, retention is slightly lower at three months (less than 1 percentage point across all ages), moderately higher at four years (up to 3 percentage points for 25 to 27 year olds, but less than a percentage point for the oldest group), and slightly higher at six years.

After correcting for observed covariates, the effect of age on promotion conditional on retention remains positive at four years—more than 4 percentage points for ages 22 to 24 and ages 25 to 27, about 1 percentage point for the oldest group. However, promotion effects at six years are negative for ages 25 to 27 (about 1 percentage point) and for ages 28 to 42 (about 4 percentage points). For unconditional promotion rates, the higher retention rates more than offset the sometimes lower promotion rates, so that total promotion rates (unconditional on retention) are usually positive and above 1 percentage point, except for the oldest recruits at the six-year mark.

CHAPTER THREE

Why Do Older Youths Join the Military?

The traditional economic models of enlistment, both theoretical and econometric, are based upon the economic theory of occupational choice (McFadden, 1983) as applied to the modern all-volunteer force (Fechter, 1970). The economic model posits that an individual considers the military and his or her best civilian alternative as two mutually exclusive choices.[1] Generally, these models assume that the decision to join or not to join is made once; that is, these are single-period models. We have seen, however, the path into the military is not as simple as such a model implies. In reality, those who graduate from high school and do not join the military can revisit their decision. If qualified, they can join the military at any point up to an age limit set by policy, and often even beyond, if a waiver to the policy is granted.

To better understand the phenomena of why a person might put off his or her enlistment in the military and of why older youths would then enlist, we developed a new conceptual economic model of youth behavior. In addition, with the cooperation of the U.S. Army Enlistment Command and the U.S. Army Training and Doctrine Command, we collected demographic and socioeconomic data from more than 5,000 new recruits, about 20 percent of whom joined the Army directly out of high school and 80 percent of whom joined later, to see how these groups might differ.

A Simple Two-Period Model of Military Enlistment

Our model builds on the traditional model that has only one period, graduation from high school, at which point a person has one opportunity to decide upon his or her choice of occupation—military service or a private sector job. A slightly more realistic model is a two-period model in which an individual can choose between military service in an active or reserve com-

[1] This is often translated into an econometric model in which a supply function is estimated, as in Asch et al., 2010, p. 15:

> Individuals are assumed to choose to enlist if the military provides greater utility or satisfaction than the best civilian alternative. Factors affecting utility include the taste for military service versus civilian opportunities; the relative financial returns to military and civilian opportunities, as well as such random factors as health or economic shocks. We do not estimate a structural model of enlistment. . . . Instead, we estimate what is known as a reduced-form model that posits that high-quality enlistments are associated with a set of variables that capture taste for military service, such as demographic factors, the financial return to military service and civilian opportunities (such as enlistment bonuses and pay) and factors that can affect the taste for military service, such as recruiters.

ponent and the private sector immediately at graduating from high school but can revisit this decision later.[2] Figure 3.1 presents a graphic representation of such a model.[3]

For a person graduating from high school, the military path is well known. Recruiters tell the prospective recruit what to expect, how he or she might advance, how much pay he or she will receive, and other related information. On the other hand, the civilian option, particularly if the high school graduate is not going to college full time, is less certain. A high school graduate does not know if he or she will be able to find employment, what wage he or she will receive, or what benefits will be. This uncertainty will drive some to join the military—take path V_M, as delineated in our model. The opportunity to explore the civilian employment alternative, with the knowledge that the military option will be there in the future, moves others to "test the market" rather than enlist immediately after graduation—path V_C. Only those who are risk averse and want a sure thing or are sure that they want the military lifestyle will enlist out of high school—path V_M. Others, even those who are inclined toward military service, can afford to take a "wait and see" attitude knowing that, if things do not work out in the civilian sector, they can always fall back on the military—path V_{CM}.

In the second period, someone who decided to try the private sector and not join the military after graduation has now gained information that will help inform the decision to stay in the private sector or join the military in the next period. For the purpose of this analysis, we assumed that the person who joined the military has no option but to continue to fulfill his contract and serve—V_{MM}. In the second period, with the information gained in the first

Figure 3.1
Simple Two-Period Model of the Decision to Enlist After High School

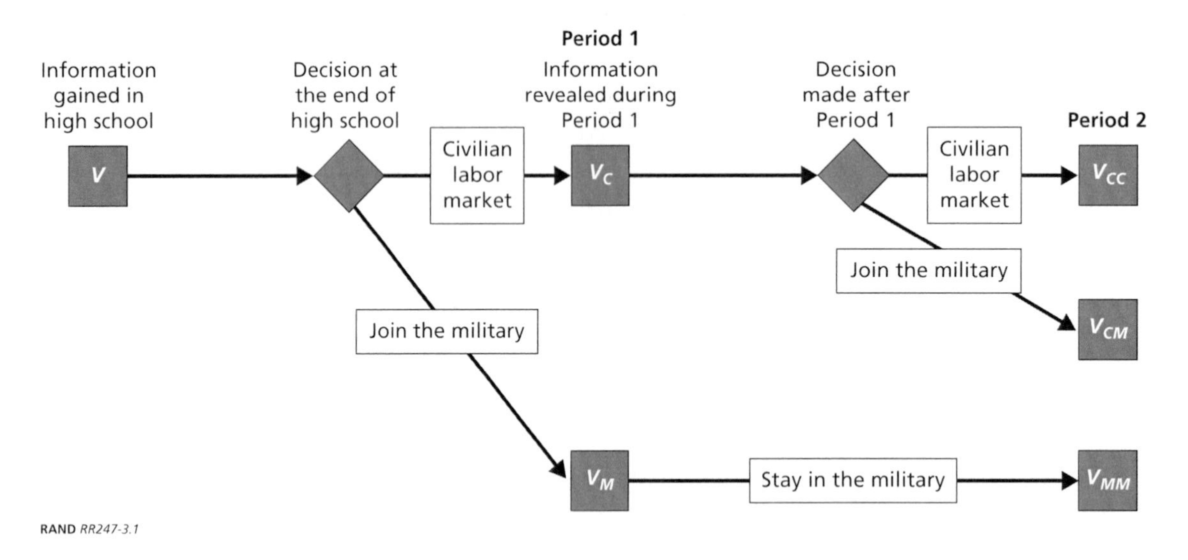

RAND *RR247-3.1*

[2] In the real world, the only fixed decision point is what to do when one graduates from high school. Those who did not join the military are free to reconsider their decision at any time. We present it here as a definition point that marks the boundary between the first and second periods only.

[3] More formally, this is a dynamic programming problem. A similar model can be developed that considers the options of staying a civilian, join a regular military component, or joining a reserve component. The last contains elements of both the civilian and military options. While the active duty and civilian options are mutually exclusive, the option of joining the reserves combines the two. Those who select the reserve option have what is under normal circumstances a full-time civilian job but are required to serve with a reserve component for training, with some probability of being activated for an extended period.

period, our subject must make another decision. Someone who is comfortable with his life in the private sector will remain a civilian—V_{CC}. If, on the other hand, things have not gone as well in the private sector as the person had hoped or if his or her basic interest in a military life style has changed, he or she might decide to join the military—V_{CM}.

Given this two-period model, we would like to know whether those who chose a particular path differ significantly from those who took a different path and what factors might have influenced the path they did take. To obtain information on each group, we surveyed Army soldiers in a way that gave us information about those who enlisted immediately after high school (V_{MM}) and those who enlisted later (V_{CM}) and compared them to each other. We then compared our sample (weighted based on key characteristics, to permit a more appropriate comparison) with youth from a national sample—National Longitudinal Survey of Youth 1997 (NLSY97)—the overwhelming majority of whom did not join the Army (V_{CC}).[4]

[4] For more on NLSY97, see Bureau of Labor Statistics, 2013.

CHAPTER FOUR

Surveys of Army Enlistees and the American Youth Population

To compare Army recruits who joined soon after leaving high school with those who joined later, RAND developed a new survey instrument that was administered to Army recruits during basic training. For all respondents, the survey recorded demographic information, family associations with the military, perceptions about recruiting activities in high school, and their reasons for joining or not joining after high school. For those who did not join the Army directly after high school, additional information was collected, including an extensive socioeconomic history of their situation after high school and preenlistment, their accounts of what had changed, and their reasons for finally deciding to join either an active or reserve component. We then compared these groups, weighted to create a more reasonable comparison, to a nationally representative sample of youth from NLSY97.

Taken together, the two surveys allowed us to describe the overall recruiting environment—both the recruiting environment while prospective enlistees are still in high school (family situations, knowledge of military and nonmilitary employment options, and reasons for joining or not joining) and the recruiting environment after high school for those who did not join at graduation. We were able to compare those who joined soon after high school—"early"—with those who joined later—"late." We were also able to compare those who joined an active duty component with those who joined a reserve component.

Survey of Army Recruits

Overview

During spring and fall 2008, we surveyed approximately 5,000 new Army recruits at each of the five Basic Combat Training (BCT) and One-Station Unit Training (OSUT) bases that the U.S. Army currently operates.[1] The bases and their locations are shown in Figure 4.1.

[1] BCT generally lasts nine or ten weeks. At the time of our surveys, three of the five training bases (Fort Benning, Fort Leonard Wood, and Fort Knox) were running nine-week BCT, which has typically been the Army standard. Two of the five bases (Fort Jackson and Fort Sill) were running experimental ten-week BCT at the recent request of the Army Training and Doctrine Command (TRADOC) to see whether an additional week of training prepared soldiers better (per TRADOC's evaluation).

OSUT combines BCT and Advanced Individual Training, always starting with the standard nine- or ten-week schedule. The length and schedule of subsequent activities in OSUT depend on the military occupational specialty (MOS). Six MOS categories are assigned to OSUT: infantry, armor, combat engineering, military police, chemical, and artillery. The OSUT training for these six MOS categories is spread across four of the five U.S. Army training bases. One base, Fort Jackson, trains soldiers only in standard BCT.

Figure 4.1
Location of U.S. Army Training Bases

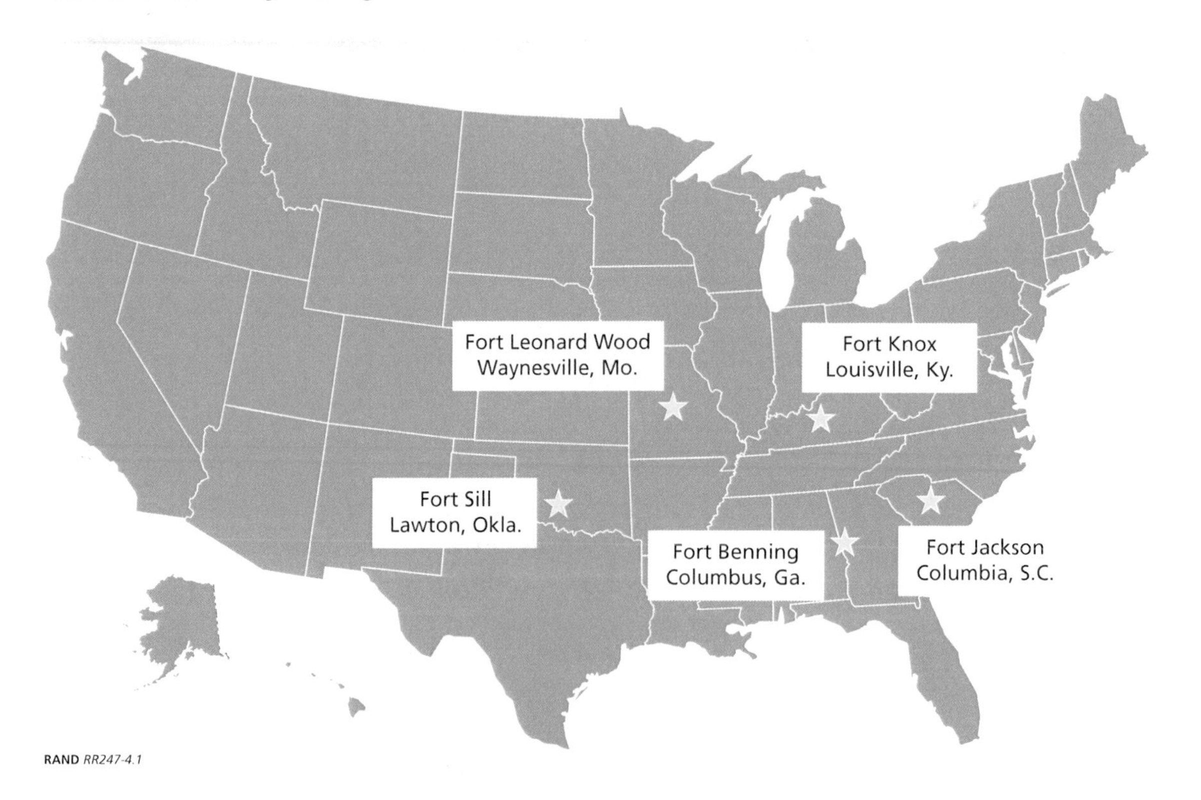

RAND *RR247-4.1*

Generally, the command at the BCT asked us to administer our survey during the last week before graduation. Administratively this was the closest time to when enlistees had made the decision to join the Army that a group survey could be given. Since we were particularly interested in the decisions of those who had not joined at high school graduation but who had joined later, e.g., those age 20 and above, (V_{CM}), we selected a time of year when we knew this group would be well represented at BCT, late winter and spring 2008. These BCT classes also included some high school graduates, and we increased the number of respondents from this group—increased the number of V_{MM}—by revisiting a number of BTC bases in fall 2008. Aside from the bias we introduced with this selection of time for our survey collection activities and the oversampling of those who joined sometime after graduation from high school, the sample was representative of recruits who joined the Army in 2008.

Organizing the Data-Collection Effort

Obtaining cooperation from senior Army officials was critical for this project. This included support from TRADOC, as well as the senior leadership at each training base (i.e., the commanding general). The project sponsor and the principal investigators traveled to TRADOC headquarters at Fort Monroe to seek its support. To work out the details of collecting the data, one of the principal investigators and the RAND survey coordinator also met with senior leadership at Fort Jackson. This included a face-to-face meeting with the commanding general to explain the study and what would be required of base personnel.

Assignment to a specific BCT or OSUT center depends on a soldier's MOS, which is determined by the recruit's AFQT score, Cognitive Ability Test score, MOS availability, and personal preference.

After completing the approval process,[2] TRADOC issued a "tasker" to each data collection site. The tasker was a one-page document that listed a Department of Defense contact for the project and asked each base to submit to the RAND survey coordinator the name and contact information for a point of contact (POC) for the study. Once the RAND survey coordinator received the POC name and contact information from a base, she worked directly with the POC on the scheduling and logistics for the data collection.

We learned from senior base leadership that it was very important that our effort interfere as little as possible with the standard BCT/OSUT schedule. As a result, most of our data collection took place during the eighth week of training, when the recruits' daily schedule is less stringent than earlier in their training; however, there were some exceptions to this protocol. The POC at Fort Knox was able to schedule soldiers for surveys more easily when they initially arrived for training; that is, before they actually began the standard BCT/OSUT schedule. At Fort Sill, some units preferred scheduling surveys for the fourth week of training instead of the eighth week because of the logistics associated with scheduling graduation ceremonies. Regardless of the variations in the timing of survey administration, we do not believe that the survey responses we received were affected; the questions in our survey focused solely on experiences prior to arrival on base for training. We do note that conducting the majority of our data collection after some training was completed means that we have no data from recruits who dropped out before our survey was administered.

While we were particularly interested in data from new recruits who were 20 years old or older (what we term "older recruits"), because they were the primary focus of our study, we also collected data from younger Army recruits to be used as a comparison (control) group. Collecting data from all recruits regardless of age had two major benefits. First, this strategy gave the analysis team an opportunity to examine differences in responses between younger and older Army recruits. As discussed later in this chapter, the analysis team found comparing responses from younger recruits to responses from older recruits to be so rich that we scheduled additional data collection trips to complete more surveys with younger recruits. Second, it simplified sampling in that we did not have to segregate respondents by age; we surveyed all recruits in a unit regardless of their age.

Collecting the Data

We collected data at the platoon or company level, with anywhere from 60 to 200 soldiers completing the survey in a group setting. Figure 4.2 shows one of several sessions held at Fort Benning, Georgia. Table 4.1 displays our final survey administration calendar and the total number of soldiers present during each data collection trip.

We received a total of 5,373 completed surveys across the five data collection sites.[3] All five training bases had a completion rate greater than 90 percent, and the differences in completion rates between sites are explainable. Fort Knox had the highest completion rate, which may have to do with the fact that we surveyed soldiers at that location just after their arrival. Fort Leonard Wood had the lowest completion rate, which may be explained by the fact that although we requested that soldiers be required to stay in the room while the survey was taking

---

[2] The approval process included submitting project and supporting documentation to the Human Subjects Protection Committee at RAND, the Army Research Institute, the Defense Manpower Data Center, and the Department of Defense.

[3] According to FY 2006 data from the U.S. Army, our sample size represents nearly 3 percent of the entire population of new recruits across all components—active Army, Army Reserve, or Army National Guard.

Figure 4.2
Classroom Configuration at Fort Benning

SOURCE: Bernard Rostker.
RAND *RR247-4.2*

Table 4.1
Data Collection Schedule, Calendar Year 2008

Data Collection Locations	Attendance							
	March	April	May	June	July	August	September	Total
Fort Benning			386		483			869
Fort Jackson	1,008						412	1,420
Fort Knox	379	552					208	1,139
Fort Leonard Wood			550		470			1,020
Fort Sill		216	711				213	1,140
Total	1,387	768	1,647		953		833	5,588

place, even if they opted not to complete the survey, drill sergeants for several companies allowed soldiers to leave the room if they were not taking the survey. Table 4.2 displays the breakdown of completed surveys per base.

Questionnaire Design

The appendix presents the final questionnaire. Instrument development began after the first year of project work was completed. During that year, the team compared the project research questions to the available survey data (in particular, NLSY97 and the U.S. Army Recruiting Command [USAREC] New Recruit Survey from Fiscal Year 2006 [NRS FY06], both

Table 4.2
Breakdown of Completed Surveys by Training Base

Site	Total Attendance	Total Completed Surveys	Completion Rate (percent)
Fort Benning	869	839	96.5
Fort Jackson	1,420	1,376	96.9
Fort Knox	1,139	1,133	99.5
Fort Leonard Wood	1,020	934	91.6
Fort Sill	1,140	1,091	95.7
Total	5,588	5,373	96.2

discussed in more detail below). Our review indicated that a new survey would improve our understanding of motivations for joining the military by supplementing the data available in the NLSY97, which provides a large sample of nonenlistees but lacks adequate numbers of enlistees, particularly older enlistees.

Many of the initial topic areas and specific questions were chosen to parallel the two surveys already in existence. As a control group, we used data from the NLSY97 to compare nonenlistees (not surveyed in this project) with enlistees (some surveyed in this project), so we were careful to word our survey questions as closely as possible to those in NLSY97.

In addition to considering the NLSY97 survey, we also examined the USAREC NRS FY06. USAREC had already developed wording for many questions of interest to our project, particularly about experience with recruiters. Noting that USAREC has experience surveying our target population, we took their questionnaire items, wording, and formatting into consideration. This allowed us to make meaningful comparisons and benchmark our results to theirs because the survey populations were similar. As a stakeholder in our project, it was also important for USAREC to be comfortable with our data, which would be more likely if we took the USAREC questionnaire items, wording, and formatting into account. Table 4.3 shows the questions in our survey that were worded exactly the same as matching questions in other surveys.

In developing our questions, we were mindful of the plan to combine our data with MEPCOM data for respondents who provided us with their Social Security Numbers (SSNs). We planned to do so because the MEPCOM data set provides additional demographic and administrative data, such as test scores.[4] We did, however, leave some questions in our survey that yielded data that were available from MEPCOM.[5] This hedged against possible delays in obtaining enlistment records or administrative data that might impact the analysis, allowed us to ask some questions both in the self-administered portion of the survey and in the calendar

[4] We were able to combine our survey data with MEPCOM data only for recruits who agreed to provide their SSNs during survey administration sessions (87 percent of all recruits who completed a survey). That made collecting SSNs on our survey very important. Using MEPCOM data also allowed the analysis team to expand on some survey data by including a question on race, following the NLSY97 wording, which is different from the approved wording from the Office of Management and Budget, which is what MEPCOM uses.

[5] The questions that we included on our survey even though they were also available in the MEPCOM file included birth date, place of birth, sex, race, marital status, date of last high school attendance, and highest level of education achieved.

Table 4.3
Items Formatted to Match Existing Questions

RAND Survey Item Number	Topic	Matching Survey
5	Race	NLS-Y97
7	Family members with military service history	USAREC NRS FY06
11	Highest grade of school completed	NLS-Y97
16	Contact with recruiters in high school	USAREC NRS FY06
70	Number of jobs	NLS-Y97
75, 80	Hours worked per week	NLS-Y97
76, 81	Tenure of job in months	NLS-Y97
77, 82	Pay	NLS-Y97
83–86	Monthly breakdown of employment status	NLS-Y97

NOTE: Appendix A contains the RAND survey.

portion of the survey (see the next subsection) that could facilitate completion of the complicated calendar section, and gave us information from two sources that allowed us to check each respondent's survey data for internal consistency.

Event History Calendar

To account fully for the socioeconomic events in the lives of recruits who did not enlist after high school but joined some time later, we developed an event history calendar (EHC),[6] as shown in Figure 4.3. In survey research, EHCs are commonly used in one-on-one interview sessions with a proctor, who has a respondent visually place past events on a time line or calendar (Freeman, 1988), sometimes using computer software (Belli, 2000). In our case, the EHC was administered in a group setting but with a proctor providing instructions to the entire group.

The EHC had four parts. *Landmark Events* asked if and when the respondent had received a high school diploma, if and when the respondent had received a GED, and when the respondent had signed the enlistment contract. *Post–High School Education and Training* asked about attendance at two-year or four-year colleges and other training or vocation programs (i.e., technical school). *Family Life* asked about marriages, separations, divorces, widowing, and children. *Employment History* asked about number of jobs, location of jobs, employers, hours and months worked, and pay received. Recruits were asked to report on up to two jobs held in each year since they left high school. Recruits who had held more than two jobs in a particular year were asked to report on the two jobs that they had held the longest. This section also asked NLSY97-like questions on status of employment throughout the calendar year: number of months employed; number of months not employed but looking for work; number of months not employed, not looking for work, but in school; and number of months not employed, not looking for work, and not in school.

[6] An EHC is sometimes also referred to as a "life history calendar."

Figure 4.3
Event History Calendar

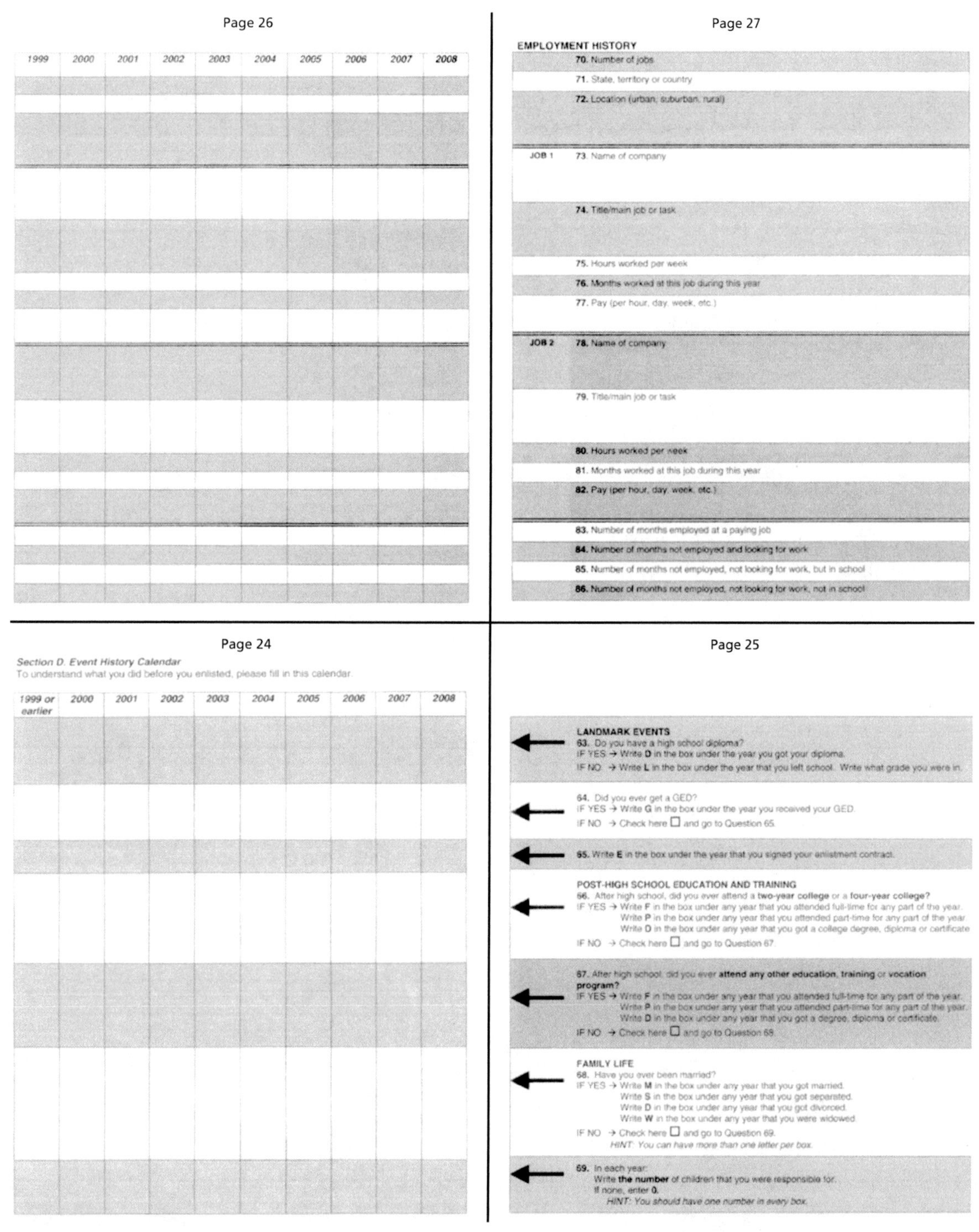
Page 26

1999	2000	2001	2002	2003	2004	2005	2006	2007	2008

Page 27

EMPLOYMENT HISTORY

70. Number of jobs
71. State, territory or country
72. Location (urban, suburban, rural)

JOB 1
73. Name of company
74. Title/main job or task
75. Hours worked per week
76. Months worked at this job during this year
77. Pay (per hour, day, week, etc.)

JOB 2
78. Name of company
79. Title/main job or task
80. Hours worked per week
81. Months worked at this job during this year
82. Pay (per hour, day, week, etc.)

83. Number of months employed at a paying job
84. Number of months not employed and looking for work
85. Number of months not employed, not looking for work, but in school
86. Number of months not employed, not looking for work, not in school

Page 24

Section D. Event History Calendar
To understand what you did before you enlisted, please fill in this calendar.

1999 or earlier	2000	2001	2002	2003	2004	2005	2006	2007	2008

Page 25

LANDMARK EVENTS
63. Do you have a high school diploma?
IF YES → Write **D** in the box under the year you got your diploma.
IF NO → Write **L** in the box under the year that you left school. Write what grade you were in.

64. Did you ever get a GED?
IF YES → Write **G** in the box under the year you received your GED.
IF NO → Check here ☐ and go to Question 65.

65. Write **E** in the box under the year that you signed your enlistment contract.

POST-HIGH SCHOOL EDUCATION AND TRAINING
66. After high school, did you ever attend a **two-year college** or a **four-year college?**
IF YES → Write **F** in the box under any year that you attended full-time for any part of the year.
Write **P** in the box under any year that you attended part-time for any part of the year.
Write **D** in the box under any year that you got a college degree, diploma or certificate.
IF NO → Check here ☐ and go to Question 67.

67. After high school, did you ever **attend any other education, training** or **vocation program?**
IF YES → Write **F** in the box under any year that you attended full-time for any part of the year.
Write **P** in the box under any year that you attended part-time for any part of the year.
Write **D** in the box under any year that you got a degree, diploma or certificate.
IF NO → Check here ☐ and go to Question 68.

FAMILY LIFE
68. Have you ever been married?
IF YES → Write **M** in the box under any year that you got married.
Write **S** in the box under any year that you got separated.
Write **D** in the box under any year that you got divorced.
Write **W** in the box under any year that you were widowed.
IF NO → Check here ☐ and go to Question 69.
HINT: You can have more than one letter per box.

69. In each year:
Write **the number** of children that you were responsible for.
If none, enter **0**.
HINT: You should have one number in every box.

RAND *RR247-4.3*

Event History Calendar Outcomes

As we expected from our successful pilot testing, respondents were able to complete the EHC. As described earlier, we asked only soldiers who had waited awhile before enlisting in the Army to complete the EHC. Respondents were only asked to provide answers to questions for the years since they left high school. We found that even the oldest recruits filled out the calendars completely.

Figure 4.4 shows the Employment History section of the EHC that an older recruit completed. This recruit had been out of high school for at least ten years (it is possible that this recruit left high school more than ten years ago, but our EHC is limited to ten years). This recruit had held at least two jobs in five of the ten years and had held one job in each of the remaining years. Information is provided on all of the years of employment in the rows and columns of this EHC.

As an internal validity check, we randomly sampled a portion of surveys to determine the accuracy of the EHC data and the survey response data. Since we asked some questions in both the non-EHC portion of the survey and in the EHC portion of the survey, we were able to compare answers. In the question concerning marital status, for example, we found that respondents' answers matched 99.5 percent of the time.

National Longitudinal Survey of Youth

In use for more than four decades, the National Longitudinal Surveys are designed to gather information at multiple points in time on the labor market activities and other significant life events of several groups of men and women.[7] NLSY97 is a survey of young men and women

Figure 4.4
Ten Years of Employment History Documented on the EHC

Page 26 | Page 27

1999	2000	2001	2002	2003	2004	2005	2006	2007	2008	EMPLOYMENT HISTORY
1	1	1	2	1	2	3	3	2	1	70. Number of jobs
MI	MI	MI	MI	MI	MI	MI	MI	MI	MI	71. State, territory or country
Urban	Sub	Sub	Sub	Sub	Sub	Sub	Sub	Sub	Sub	72. Location (urban, suburban, rural)
[illegible] Builders	Self employed	Three Rivers Construction	Three Rivers Construction	Campbell Masonry	Campbell Masonry	Northland Siding	Self employed	self employed	Rudy Builders	JOB 1 73. Name of company
Carpentry	Carpentry	Carpentry	Carpentry	masonry	Masonry	Carpentry	Carpentry	Carpentry	Carpentry	74. Title/main job or task
30	35	40	40	45	45	35	50	50	40	75. Hours worked per week
12	12	12	5	10	3	4	6	4	1	76. Months worked at this job during this year
18 pr. hr.	$500 pr. wk.	12 pr. hr	13 pr. hr.	12 pr. hr	12 pr. hr.	12 pr. hr.	700 per wk	700 pr. wk.	15 per hr.	77. Pay (per hour, day, week, etc.)
			Campbell Masonry		Northland Siding	Campbell Masonry	Northland Siding	Rudy Builders		JOB 2 78. Name of company
			Masonry		Carpentry	Masonry	Carpentry	Carpentry		79. Title/main job or task
			40		35	35	40	40		80. Hours worked per week
			5		7	5	5	6		81. Months worked at this job during this year
			11 pr. hr.		12 pr. hr.	12 pr. hr.	15 pr. hr.	15 pr. hr.		82. Pay (per hour, day, week, etc.)
12	12	12	10	10	10	9	11	10	1	83. Number of months employed at a paying job
0	0	0	2	2	2	3	1	2	0	84. Number of months not employed and looking for work
0	0	0	0	0	0	0	0	0	0	85. Number of months not employed, not looking for work, but in school
0	0	0	0	0	0	0	0	0	0	86. Number of months not employed, not looking for work, not in school

RAND *RR247-4.4*

[7] Information for this section is derived liberally from the National Longitudinal Program, 2011.

born in the years 1980 through 1984. It is a nationally representative sample of approximately 9,000 individuals who were 12 to 16 years old as of December 31, 1996.

The NLSY97 is designed to document the transition from school to work and into adulthood. It collects extensive information about youths' labor market behavior and educational experiences over time. Employment information focuses on two types of jobs, "employee" jobs, working for a particular employer, and "freelance" jobs, such as lawn mowing and babysitting. These distinctions enable researchers to study the effects of very early employment among youths. Employment data include start and stop dates of jobs, hours worked, and earnings. Measures of work experience, tenure with an employer, and employer transitions can thus be calculated. Educational data include youths' schooling history, performance on standardized tests, courses of study, the timing and types of degrees, and detailed accounts of progression through postsecondary schooling.

The first survey (round 1) took place in 1997. In that round, each eligible youth and one of his or her parents received hour-long personal interviews. In addition, during the screening process, an extensive two-part questionnaire listed and gathered demographic information on members of the youth's household and on his or her immediate family members living elsewhere. Youths were thereafter interviewed annually. In 1997 and early 1998, the NLSY97 respondents were given the computer-adaptive version of the ASVAB, which comprises ten tests that measure knowledge and skill in a number of areas, including mathematics.

Comparison of Samples

Our sample is generally similar to the profile of Army recruits obtained from the MEPCOM, as shown in Table 4.4. AFQT categories I and II were overrepresented when compared with MEPCOM data, and category I was slightly underrepresented when compared to the NLSY97. The NLSY97 reports that almost 30 percent of the youth population scores at or below an AFQT category IV. The military takes very few category IV personnel. Given this policy,

Table 4.4
Comparison of Sample Characteristics

AFQT Categories	NLSY97	MEPCOM	RAND Survey	
			Unweighted	Weighted[a]
I	7.54	4.71	6.23	6.34
II	28.95	30.56	33.24	33.98
IIIA	14.95	24.69	23.61	23.29
IIIB	19.21	37.10	33.80	33.40
Less than IIIB	29.35	2.94	3.12	2.99
Male	49.99	81.16	87.41	83.84
Black	15.64	14.38	14.94	16.16

[a] We computed the joint distribution of our weighted survey for three variables: gender, race (white or black), and AFQT category. We then reweighted the AFQT data to represent that same distribution. Using the new weights, we computed the revised weight used in this table.

the numbers of category II and III recruits make up a larger percentage of total Army enlistments than of the general population. While today's military employs more women than at the advent of the all-volunteer force in 1993, it is still dominated by male recruits. Black representation appears roughly equal between the three sets of data.

CHAPTER FIVE

What We Learned About Older Recruits: An Analysis of Survey Results

This chapter describes the results of the RAND survey of older recruits. We begin our discussion with results obtained from the Army survey data and conclude with observations about how the results from our survey of older recruits, weighted by key characteristics to produce a more comparable sample, compare with American youth from the NLSY97. Note that, in many cases, survey respondents could select more than one answer; therefore, the percentages presented below often add up to more than 100 percent.

The Military Is a Family Business

Our survey data suggest that the military has become a family business. In our total sample of those who joined early and late, regardless of the component they joined, most had a close family association with the military. Eighty-three percent of our survey respondents had a close family member who had served in the military. As shown in Figure 5.1, these numbers did not

Figure 5.1
Many Close Family Members Have Served

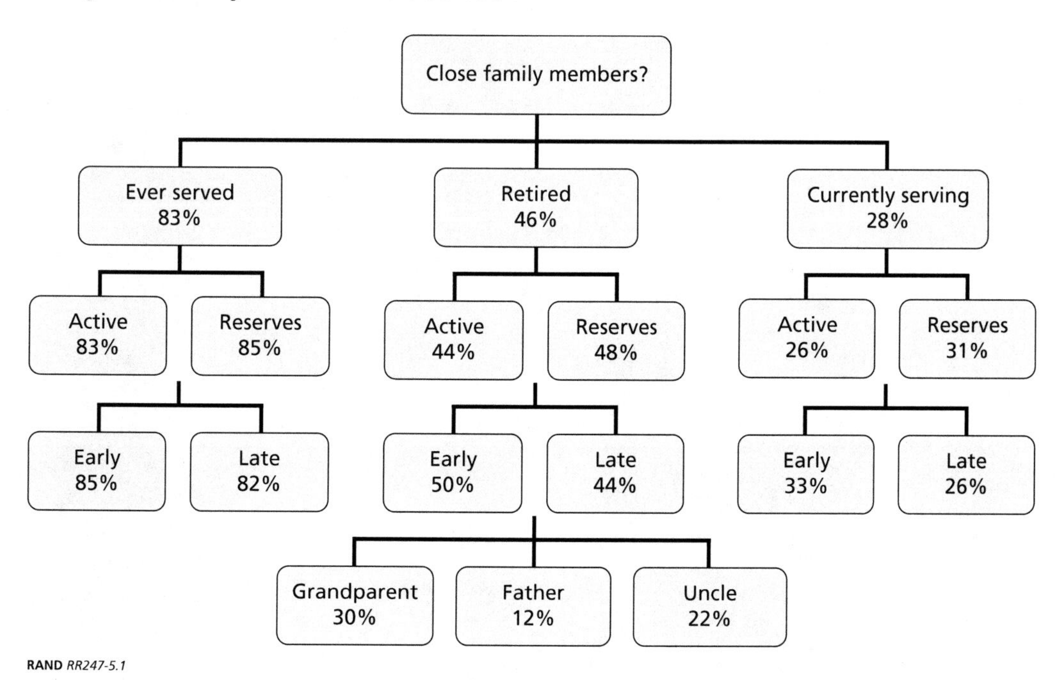

RAND *RR247-5.1*

differ much for those who enlisted to serve in an active or reserve component or who enlisted soon after high school or some time later. Even more impressive was the fact that almost one-half of our sample had a close family member that had retired from the military, one-third of whom were grandparents and almost a quarter of whom were uncles. The lower figure for fathers reflects the fact that the parents of some recruits were still serving when the survey was taken.

We asked the recruits which of their relatives had ever served (Figure 5.2). We were particularly interested in the percentage of the survey respondents who had fathers and mothers serving in the military, since a comparable national statistic is available from the Department of Labor's Current Population Survey (CPS). As noted, the percentage of recruits with fathers (38 percent) or mothers (6 percent) who served in the military is many times greater (and the difference is statistically significant) for our sample than for the total U.S. population.

Encouragement from parents, relatives, and friends was important, especially for those who joined immediately after high school. As shown in Figure 5.3, these recruits were much less frequently encouraged to join the military by boy- or girlfriends.

Figure 5.2
Among Recruits, Service by Fathers and Mothers Is Prevalent

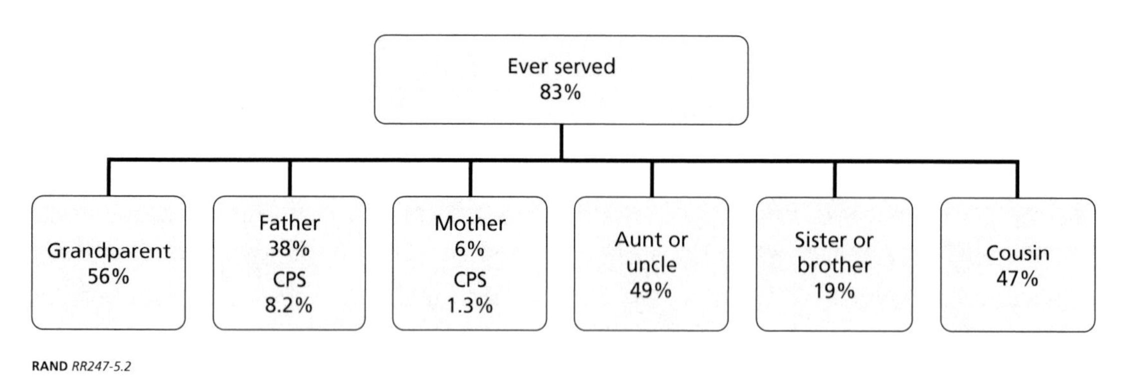

RAND *RR247-5.2*

Figure 5.3
Parental Support Was Greater for Those Who Joined Out of High School

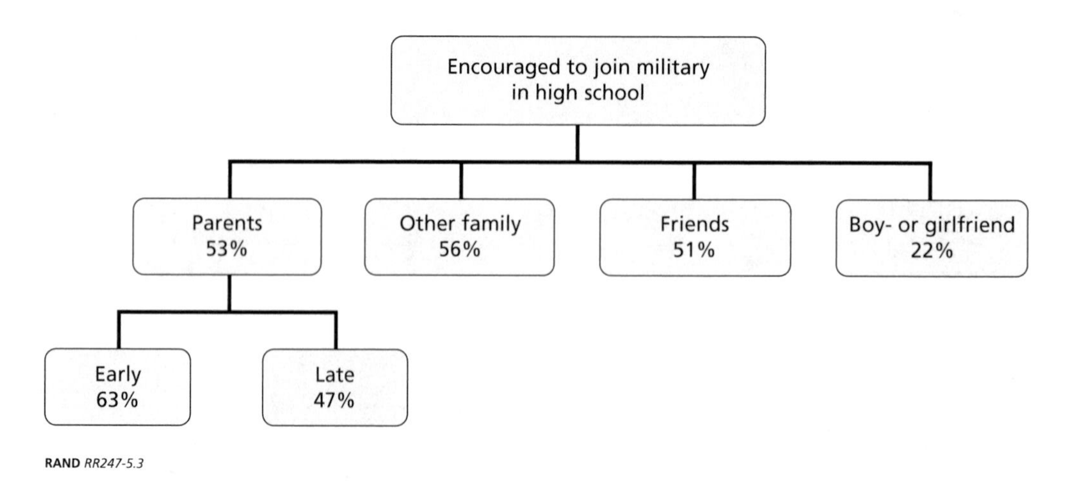

RAND *RR247-5.3*

Recruiting

The high school has been a central focus of recruiting since the advent of the all-volunteer force in the 1970s. The respondents to our survey overwhelmingly reported that recruiters had come to their high schools—73 percent. As reported in Figure 5.4, slightly more of those who enlisted in reserve components reported that recruiters had visited their high schools than those who joined an active duty component. The difference is not statistically significant. There is, however, a significant difference with respect to when respondents enlisted. We divided the time of enlistment into three groups relative to the time respondents left high school. Significantly more respondents who enlisted soon after graduation reported that recruiter had visited their high schools. Clearly, the role of recruiters warrants further study because of the possibility of reverse causality: Those who joined early could have been more likely to pay attention to visiting recruiters.

In addition to learning about military opportunities from recruiters, recruits used a variety of other sources, including traditional and new media outlets. As shown in Figure 5.5, 87 percent of recruits used the Internet to learn about the military; most frequently, they used the services' websites. Pop-up ads were also important. Relatively few of our respondents reported that they learned about the military from online games. Traditional media also remained important. Many recruits learned about military opportunities through television, magazines, and news stories. Fewer recruits learned about the military by listening to the radio or by reading newspapers or books.

However, the way older recruits interacted with the Army was very different from the way younger recruits did. Older recruits, as shown in Table 5.1, actively sought out Army recruiters. Programs built around the school were much less useful to late recruits. Only 24 percent of later recruits indicated that they had made contact through their schools, compared with 73 percent of those who enlisted after high school. The percentage who indicated that they had learned about military opportunities through postings at school was down from 34 to 16 percent; the percentage who learned about such opportunities at job fairs was down from 23 to 12 percent. These older recruits were much more likely to stop by recruiting stations and/or fill out request postcards.

Figure 5.4
In High School, Most Students Learned About the Military from Recruiters

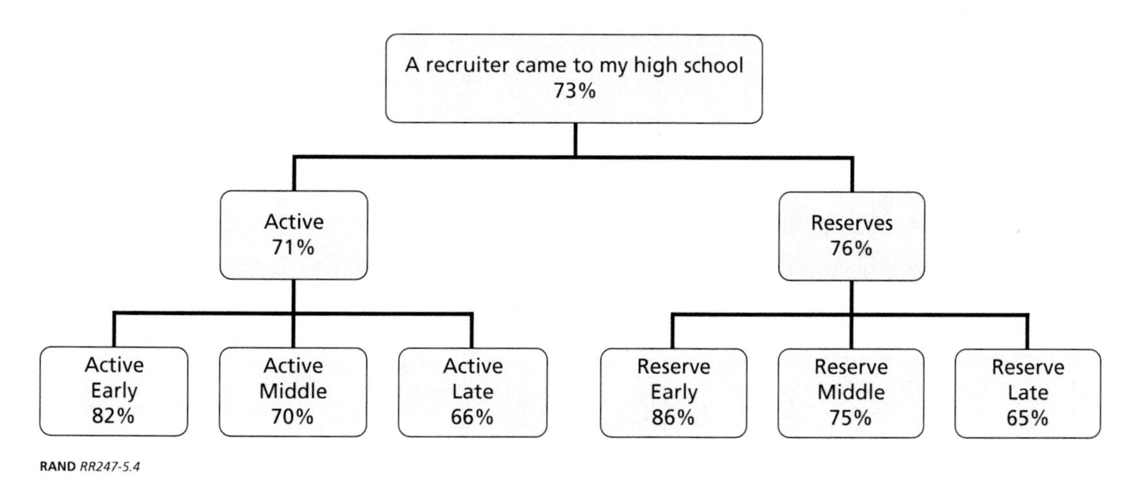

RAND *RR247-5.4*

Figure 5.5
High School Students Also Learned About the Military from a Variety of Other Sources

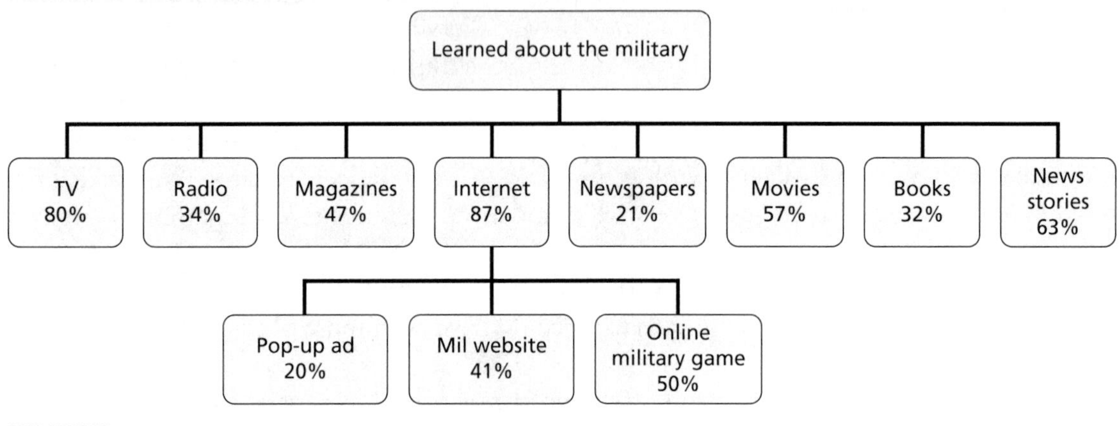

RAND *RR247-5.5*

Table 5.1
Contact with the Army Differs for Early and Late Joiners (percent)

Factor	From High School "Early Joiners"	Later Enlistment "Late Joiners"
At school	73	24
Introduced to recruiter by friend	26	27
Stopped by recruiting station	40	77
Filled out request card/phoned	29	64
Chance encounter	Not Asked	14
Military event	13	12
Job fair	23	12
Posted at school	34	16

When Recruits Joined and Why

Why Did Some Wait to Join the Military?

As discussed in the introductory section, more than one-half of Army recruits do not join immediately after high school. Some decided to continue their education; about 55 percent of our sample of "late enlistees" indicated that they had gone to college and/or vocational school after graduating from high school; some, about 38 percent, took time off. As shown in Figure 5.6, the overwhelming number went to work. Of those who joined later, one-quarter indicated that the main reason for not joining after high school was either that someone did not want them to enlist or that they were concerned about the war. These future recruits eventually put concern about the war aside, and the views of others changed or were no longer as important when they decided to enlist.

Why Did Those Who Waited Choose to Join?

We were particularly interested in the reason that those who did not immediately join the military eventually decided to join. About one-third of those who joined later said there were "no jobs at home." There was some difference between those who joined the regular Army

Figure 5.6
There Were a Number of Reasons Students Waited to Join the Army

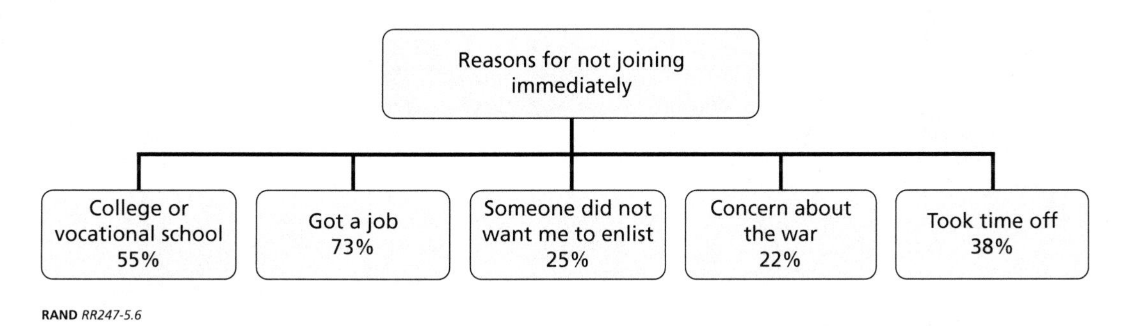

RAND *RR247-5.6*

and those who joined one of the Army's reserve components. As one might expect, those who selected active service were considerably more pessimistic than those who enlisted in a reserve component. Enlisting in the regular Army takes the recruit from his or her hometown. Enlisting in a reserve component generally means that the new service member will stay in his or her hometown and may keep his or her existing job. Forty percent of the recruits who said there were "no jobs at home" joined the regular Army, compared to 31 percent who joined a reserve component (Figure 5.7). More prevalent than these, at almost 50 percent, were those who held the view that there were only "dead-end jobs" at home, as shown in Figure 5.8. Again, those who joined the regular Army were more pessimistic than those who enlisted in a reserve component. Since these are all soldiers who "waited a while before enlisting," we thought it important to distinguish how long they waited, e.g., two or less years ("middle") or greater than two years ("late").

There were some differences between the reasons that older recruits gave for enlisting and those who enlisted right after graduation from high school gave, but in general, responses were similar, as shown in Table 5.2. The largest difference regards the importance of benefits. The discussions we had with soldiers as we were putting this survey together indicated that the erosion of health and retirement benefits in the private sector stood out in their minds. These individuals looked to employment with the Army as a way of protecting their futures. This

Figure 5.7
Among Those Who Enlisted Late, Many Thought There Were No Jobs at Home

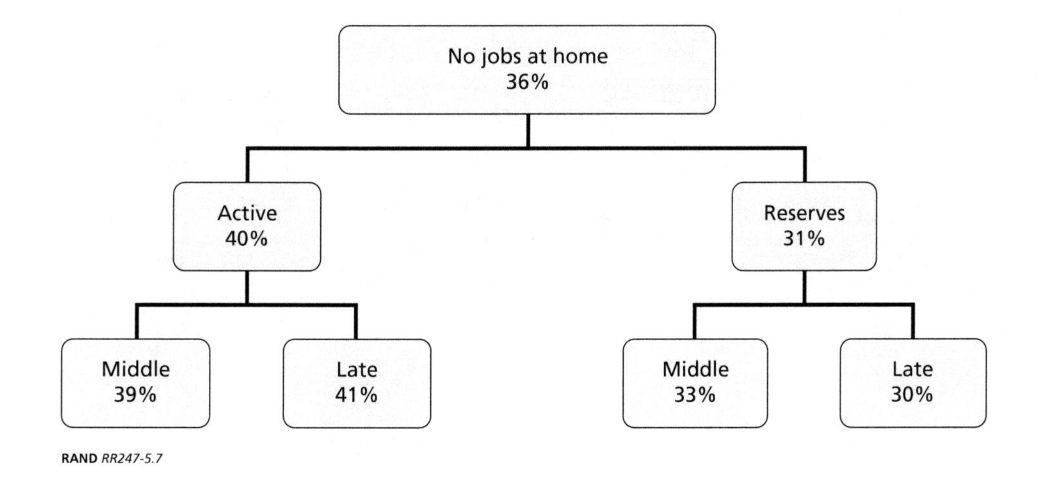

RAND *RR247-5.7*

Figure 5.8
Among Those Who Enlisted Late, Many Thought There Were Only Dead-End Jobs at Home

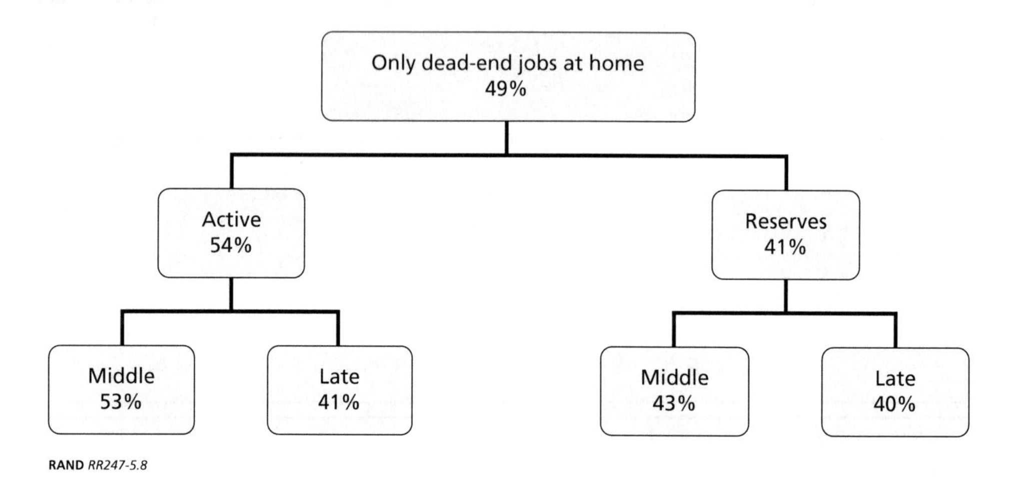

RAND *RR247-5.8*

Table 5.2
Those Who Enlisted Late Had Somewhat Different Motives from Those Who Enlisted Early (percent)

Factor	In High School "Early Joiners"	Later Enlistment "Late Joiners"
Desire to get away	61	58
No job at home	29	36
Dead-end jobs at home	40	49
Patriotism	74	71
Change of life	80	85
Money for Education	80	87
Bonus	71	79
Benefits	78	89
Pay	70	72

might become clearer in follow-up work looking at retention differences between those who had private-sector experience—late joiners—and those who made their retention decisions without the benefit of working in the private sector—those who joined right after high school.

For those who said they did not join after high school because "someone did not want me to join," Figure 5.9 shows how this played out for the older recruits. Some of those who had influenced the earlier decision changed their minds; in other cases, recruits indicated that the opinions of those individuals no longer mattered or that those individuals no longer had an influence on their decision. For this group, parental support for the decision to join the Army increased slightly over time, as shown in Figure 5.10.

Figure 5.9
The Role of Influencers Changed

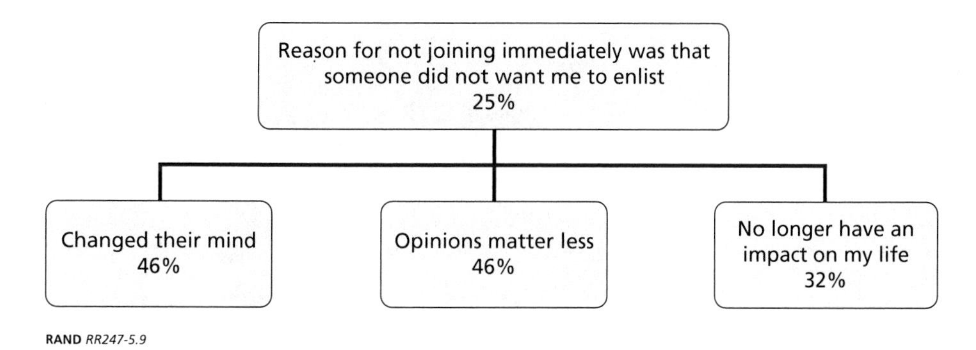

RAND *RR247-5.9*

Figure 5.10
For Those Who Enlisted Late, Parental Support Increased Slightly Over Time

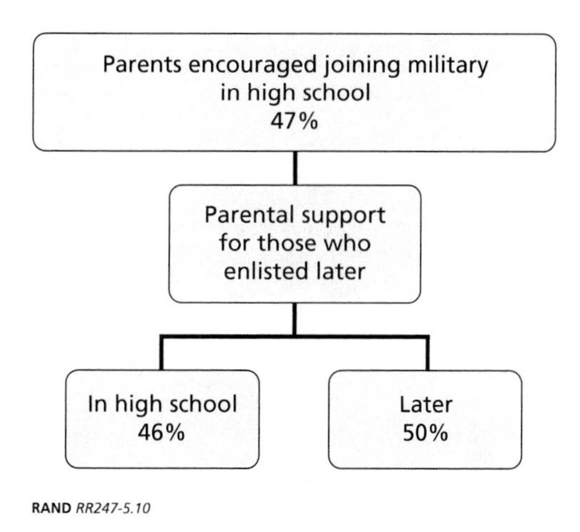

RAND *RR247-5.10*

How Do Those Who Enlist Late Compare with a Nationally Representative Cohort of American Youth?

We weighted our survey results for the group of recruits who joined after high school, as discussed previously, and compared the results with characteristics of a nationally representative group of American youth from NLSY97, most of whom did not join the Army after high school. In general, we found that those who joined the Army had not done as well since leaving high school as the average American youth in this group NLSY97. As shown in Table 5.3, they were significantly less likely to have attended a two- or four-year college. They may have been more likely to attend a two-year program as seen by their postsecondary education graduation rates in the second and third years after high school. However, in the fourth and fifth year after high school, when those attending four-year colleges received their degrees, the graduation rate for the recruits was substantially below the general youth cohort. The Army group included many more high school dropouts and very many more who had enrolled in and passed the GED examination to receive their high school diplomas after their high school classes had graduated. Consistent with the theory discussed in the previous section, the Army group worked less than the general cohort.

Table 5.3
Those Who Enlisted Late Performed Less Well than a Nationally Representative Cohort of Americans (percent)

Years After High School	Group	2- or 4-year College			High School Graduate	GED Diploma		Worked This Year
		Attending This Year	Ever Attended	Ever Graduated		Passed This Year	Ever Passed	
2	NLSY	58.14	59.63	0.13	83.99	2.29	4.79	87.79
	Army	34.85	46.49	1.07	78.78	11.41	20.48	66.92
3	NLSY	54.92	64.54	1.67	85.32	1.44	6.24	87.79
	Army	23.08	45.50	1.76	68.78	13.53	31.05	68.73
4	NLSY	52.63	68.45	4.85	89.00	0.09	7.17	87.07
	Army	19.29	48.33	3.69	70.41	8.98	29.82	69.51
5	NLSY	50.07	72.16	15.65	88.77	0.48	7.64	87.74
	Army	16.97	54.50	7.57	75.59	7.05	25.80	67.47

SOURCE: NLSY97 Cohort versus those who enlisted in the army by year since high school.

Comparing the group of Army recruits we surveyed who had decided not to join the Army after high school but who joined later against similar American youth reveals a group doing less well than other youth. Our findings are consistent with the model's suggestion that these older recruits had tested the world of work and found it wanting. Few went to college. A larger number were high school dropouts, but they had studied for and passed the GED examination when they found that they needed a high school diploma to join the Army. They had worked less than the average youth. For these young Americans, the Army provided a second chance. For those who joined the regular Army, this was a chance to leave home and start over again. They overwhelmingly indicated that they understood that they were likely to be assigned to a combat zone, but obviously, this did not dissuade them from seeking out the Army and joining.

CHAPTER SIX

Conclusions and Recommendations

Conclusions

The survey of Army recruits we conducted for this study provided insights into older recruits—those who do not join the military directly after high school graduation but, after embarking on a different path, join the military later. We learned several things about these recruits:

- More than one-half of the Army's recruits do not join immediately after high school graduation.
- Both older recruits and those who join early had close family associations with the military.
- Older recruits rely less on school-related resources to connect with the military. They are self-motivated and far more likely to stop by recruiting stations and/or fill out request postcards.
- Older recruits favored entering the job market over entering the military ate high school graduation, although many also sought to further their education.
- Older recruits who eventually joined the army often did so because they lacked work or saw only dead-end jobs in their future.
- In a comparison of a weighted sample of these recruits to a nationally representative sample of American youth, the recruits tended to have performed less well in postsecondary education and had a poorer experience in the world of work.
- Once enlisted, older recruits tend to perform as well as or better than younger recruits. In general, they have higher retention and promotion rates.

These results suggest that the older youth market will continue to be a valuable source of future recruits. To tap into this market and better understand their performance in the military, we suggest two areas for further study in our recommendations.

Recommendations

The Army may want to invest some of its recruiting resources on developing programs targeted at older youth who do not go to college. This is a very large pool of potential recruits to which little attention has been paid. It already makes up a significant and increasing portion of total enlistments, typically without being the focus of any significant recruiting effort. Currently, recruiting resources focus on high school students about to graduate—admittedly, still the most important single cohort for recruiting. Furthermore, after joining the military, the performance of older recruits is quite good, in many cases outpacing their younger colleagues who entered military service directly out of high school. While this recruiting market has tremendous potential, both in terms of its size and its history of good performance once in service, few if any existing recruiting programs are designed to tap into this market. Whether a dedicated program that focuses on these potential recruits might be able to expand the current market is an unanswered question. Older recruits are largely self-motivated, tending to seek out recruiters on their own; however, some emphasis on developing strategies to reach out to them could be fruitful. The question of how to penetrate this market was beyond the scope of the research conducted here but deserves further consideration.

The RAND survey produced a rich data set that contains unique information about the decision to join the Army and the life experiences of older recruits between the time they left high school and when they enlisted that can be the starting point for a number of potentially valuable analyses. The most logical initial question is to examine how well this sample of recruits performed during the course of their careers compared to those who joined after high school. A follow-up study to see how many completed their first terms of service, how many reenlisted, and at what rates they were promoted could answer that question. Such analysis could be accomplished using administrative records, with a relatively small investment in resources. Although we did document that older recruits, in general, tend to have higher rates of retention and promotion than younger recruits using MEPCOM data, following the individuals surveyed would allow a richer examination of the relationship between attitudes, civilian alternatives, and performance in the military. More-robust analyses of the military experience of older recruits could also be performed, with additional data collected through a follow-up survey. It would be possible, for example, to analyze the relationship between pre-service employment and employment in the military—an understanding of which could be useful in steering these youth to successful career fields.

APPENDIX

The Survey

The following pages replicate the survey on which this report is based.

RCS# DD-P&R(OT)2305
Expires January 31, 2011

RAND RECRUITING SURVEY

January 24, 2008

PRIVACY ACT STATEMENT

AUTHORITY: 5 U.S.C. 7101 Note, Employee Surveys (Pub.L. 108-136, Sec. 1128); 10 U.S.C. 136, Under Secretary of Defense for Personnel and Readiness; 10 U.S.C. 481, Racial and Ethnic Issues; Gender Issues: Surveys; 10 U.S.C. 1782, Surveys of Military Families; 10 U.S.C. 2358, Research and Development Projects; DoD Directive 5124.2, Under Secretary of Defense for Personnel and Readiness (USD(P&R)); and E.O. 9397 (SSN). This collection is conducted under Office of Secretary of Defense System of Record Notice DMDC 08, Survey and Census Data Base.

PRINCIPAL PURPOSE: The purposes of the system are to sample attitudes toward enlistment and determine reasons for enlistment decisions. This information is used to support manpower research sponsored by the Department of Defense and the military services.

ROUTINE USES: The information may be used to support manpower research sponsored by other Federal agencies.

DISCLOSURE: Voluntary. RAND is working on conducting a study for the Department of Defense to learn about why Army recruits enlist. The information collected will be used to help improve recruiting programs and policies. Information shared will not be released and will be treated as confidential.

INSTRUCTIONS

1. Do not start until you get further instructions.
2. Fill out the next page (Information Sheet) if you agree to take part in the attached survey.
3. Read and follow the instructions carefully.
4. Answer the questions by placing an X in the box next to your answer or filling in the answer as instructed.
5. If you aren't sure how to answer a question, raise your hand.
6. If you make a mistake, erase it or put a line through it and mark the correct answer.

RAND Recruiting Survey

Page 2

INFORMATION SHEET

REMINDER: It is up to you to decide if you want to take part in this activity or not. If you don't want to take part, you don't have to. If you agree to take part, you may skip any questions you don't want to answer or stop taking part at any time. We will use all information for research purposes only. RAND will treat all information you share with us as strictly confidential. We will not release any data or information that could identify you or your family to anyone.

Please complete items 1-4 if you agree to take part in the attached survey.

1. Your Name

   First Name: ______________________________ Middle Name: ___________________

   Last Name: ____

2. Your Social Security Number: ___ ___ ___ - ___ ___ - ___ ___ ___ ___

   NOTE: We will use your Social Security Number to link your survey responses to your Armed Forces Qualifying Test (AFQT) scores and possibly other administrative data. RAND will use this information to better understand the factors that might have affected recruits' decision to enlist in the Army. These data will not in any way directly affect your military career.

   If you agree to take part in the RAND Survey, but do not want to provide your Social Security Number, please leave #2 blank.

3. Today's date: Month: __________ Day: _________ Year: _________

4. Your Signature: ____

5. If you agree to take part and you would like to receive a copy of the results from this study, please print your email address below and we will email you the report as soon as it is available.

__________________________________ @ __________________________ . ______________

Name (e.g., *jdoe)* Server (e.g., *aol)* Domain (e.g., *com*)

Please remove this page from the survey booklet when you are finished.
RAND Staff will collect it when everyone has finished.

RAND Survey ID Label

Page 3

Section A. Demographic Information

1. What is your date of birth? Month: ☐☐ Day: ☐☐ Year: ☐☐☐☐ *16-23/*

2. Where were you born?

   [1]☐ United States. State: ______________________________ *24-26/*
   [2]☐ Another country. Country: ___________________________ *27-29/*

3. Are you male or female?

   [1]☐ Male *30/*
   [2]☐ Female

4. Are you of Hispanic, Latino or of Spanish descent?

   [1]☐ Yes *31/*
   [0]☐ No

5. What is your race? Mark one or more races to indicate what you consider yourself to be.

   [1]☐ White *32/*
   [2]☐ Black or African American *33/*
   [3]☐ Asian *34/*
   [4]☐ Native Hawaiian or Other Pacific Islander *35/*
   [5]☐ American Indian or Alaska Native *36/*

6. What is your current marital status? Check one.

   [1]☐ Single, never been married *37/*
   [2]☐ Married
   [3]☐ Separated
   [4]☐ Divorced
   [5]☐ Widowed

CARD 01

Page 4

7. Have any of the following family members ever served in any of the military services? Include family members who are currently serving. Check all that apply.

[1]☐ Grandparent 38/
[2]☐ Mother/Stepmother 39/
[3]☐ Father/Stepfather 40/
[4]☐ Spouse 41/
[5]☐ Boyfriend/Girlfriend/Fiancé(e) 42/
[6]☐ Aunt/Uncle 43/
[7]☐ Sister/brother or stepsister/stepbrother 44/
[8]☐ Extended family (Cousin, In-laws) 45/
[9]☐ Other: ______________________________ 46-47/
[10]☐ None of the above 48/

8. Did any of the following family members retire from the military? Do not include family members who are currently serving. Check all that apply.

[1]☐ Grandparent 49/
[2]☐ Mother/Stepmother 50/
[3]☐ Father/Stepfather 51/
[4]☐ Spouse 52/
[5]☐ Boyfriend/Girlfriend/Fiancé(e) 53/
[6]☐ Aunt/Uncle 54/
[7]☐ Sister/brother or stepsister/stepbrother 55/
[8]☐ Extended family (Cousin, In-laws) 56/
[9]☐ Other: ______________________________ 57-58/
[10]☐ None of the above 59/

9. Are any of the following family members currently serving in any of the military services? Check all that apply.

[1]☐ Grandparent 60/
[2]☐ Mother/Stepmother 61/
[3]☐ Father/Stepfather 62/
[4]☐ Spouse 63/
[5]☐ Boyfriend/Girlfriend/Fiancé(e) 64/
[6]☐ Aunt/Uncle 65/
[7]☐ Sister/brother or stepsister/stepbrother 66/
[8]☐ Extended family (Cousin, In-laws) 67/
[9]☐ Other: ______________________________ 68-69/
[10]☐ None of the above 70/

CARD 01

Page 5

10. In what month and year did you last attend high school? Month: ☐☐ Year: ☐☐☐☐

71-76/

11. What is the highest grade or level of school that you have completed? If you were home-schooled, check here ☐ and select the equivalent grade below. *77/*

   - 1☐ Did not graduate from high school/GED *78/*
     - ➜What grade were you in when you left school? Grade: ☐☐ *79-80/*
   - 2☐ High school graduate
   - 3☐ Attended college, but less than a semester and no degree
   - 4☐ Attended college, completed a semester or more but no degree
   - 5☐ 2-year college degree
   - 6☐ 4-year college degree
   - 7☐ More than 4-year college degree

12. As of today, what high school degree or certificate, if any, do you have? Check all that apply.

   - 1☐ No degree or certificate → Go to Question 13 on Page 6. *81/*
   - 2☐ High School diploma → Go to Question 13 on Page 6. *82/*
   - 3☐ GED → Answer Questions 12a and 12b below. *83/*

     12a. In what month and year did you get your GED? Month: ☐☐ Year: ☐☐☐☐

     84-89/

     12b. Did you get your GED through the National Guard Challenge Program?

       - 1☐ Yes *90/*
       - 0☐ No
   - 4☐ Other: ______________________________

   91-92/

CARD 01

Page 6

Section B. High School Recruitment Experience

CARD 02 6-7/ 1-5/

13. When you were in high school, did you learn about the military from any of the following people, even if you did not enlist then? Check yes or no for each one.

	Yes	No	
a. Relative	1 ☐	0 ☐	8/
b. Friend	1 ☐	0 ☐	9/
c. Teacher or guidance counselor at school	1 ☐	0 ☐	10/
d. Coach	1 ☐	0 ☐	11/
e. Clergy.	1 ☐	0 ☐	12/
f. Judge, youth counselor, or someone working in law enforcement	1 ☐	0 ☐	13/
g. Someone else	1 ☐	0 ☐	14/
→ Who? ______________			15/

14. When you were in high school, did you learn about the military from any of the following Internet sources, even if you did not enlist then? Check yes or no for each one.

	Yes	No	
a. Banner or pop-up ad on a website	1 ☐	0 ☐	16/
b. Military website	1 ☐	0 ☐	17/
c. Other website	1 ☐	0 ☐	18/
d. Blog	1 ☐	0 ☐	19/
e. Online military games.	1 ☐	0 ☐	20/
f. Email from someone other than a recruiter	1 ☐	0 ☐	21/
g. Chat room	1 ☐	0 ☐	22/

CARD 02

Page 7

15. When you were in high school, did you learn about the military from any of the following media sources, even if you did not enlist then? Check yes or no for each one.

	Yes	No	
a. TV ad	1 ☐	0 ☐	23/
b. Radio ad	1 ☐	0 ☐	24/
c. Magazine ad	1 ☐	0 ☐	25/
d. Newspaper display ad or help-wanted ad	1 ☐	0 ☐	26/
e. Movies.	1 ☐	0 ☐	27/
f. Books	1 ☐	0 ☐	28/
g. News or feature story on TV	1 ☐	0 ☐	29/

16. When you were in high school, did you have any of the following types of contact with military recruiters from any of the services, even if you did not enlist then? Check yes or no for each one.

	Yes	No	
a. A recruiter came to my high school	1 ☐	0 ☐	30/
b. A friend/relative introduced me to a recruiter	1 ☐	0 ☐	31/
c. A recruiter called me	1 ☐	0 ☐	32/
d. I filled out a request form for a recruiter to call me	1 ☐	0 ☐	33/
e. I stopped by a recruiting station.	1 ☐	0 ☐	34/
f. A recruiter sent me an email	1 ☐	0 ☐	35/

CARD 02

Page 8

17. When you were in high school, did you have any other contact with any of the military services, even if you did not enlist then? Check yes or no for each one.

	Yes	No	
a. I took the ASVAB at my high school	1 ☐	0 ☐	36/
b. I took the ASVAB somewhere other than high school	1 ☐	0 ☐	37/
c. I got a letter or a postcard in the mail	1 ☐	0 ☐	38/
d. I saw information posted at school	1 ☐	0 ☐	39/
e. I went to a military-sponsored event	1 ☐	0 ☐	40/
f. I went to a job fair or career fair	1 ☐	0 ☐	41/
g. I was in JROTC.	1 ☐	0 ☐	42/
→ Which service? ________________			43/

18. Were you interested in the military when you were in high school, even if you did not enlist then? Check one.

1☐ Yes → Continue. 44/

0☐ No → Go to Question 26 on Page 11.

19. Did any of the following opportunities for experience interest you in the military when you were in high school, even if you did not enlist then? Check yes or no for each one.

	Yes	No	
a. Skills	1 ☐	0 ☐	45/
b. Travel	1 ☐	0 ☐	46/
c. Adventure	1 ☐	0 ☐	47/
d. Work experience	1 ☐	0 ☐	48/

20. Did any of the following people encourage your interest in the military when you were in high school, even if you did not enlist then? Check yes or no for each one.

	Yes	No	
a. Parents	1 ☐	0 ☐	49/
b. Spouse	1 ☐	0 ☐	50/
c. Other family	1 ☐	0 ☐	51/
d. Boyfriend/Girlfriend/Fiancé(e)	1 ☐	0 ☐	52/
e. Friends	1 ☐	0 ☐	53/
f. Recruiter	1 ☐	0 ☐	54/

CARD 02

Page 9

21. Did any of the following benefits interest you in the military when you were in high school, even if you did not enlist then? Check yes or no for each one.

	Yes	No	
a. Money for education	1 ▯	0 ▯	55/
b. Bonus	1 ▯	0 ▯	56/
c. Benefits (like health care)	1 ▯	0 ▯	57/
d. Pay	1 ▯	0 ▯	58/

22. Did anything else interest you in the military when you were in high school, even if you did not enlist then? Check yes or no for each one.

	Yes	No	
a. Desire to get away	1 ▯	0 ▯	59/
b. No jobs at home	1 ▯	0 ▯	60/
c. Only dead end jobs at home	1 ▯	0 ▯	61/
d. Patriotism/Serve my country	1 ▯	0 ▯	62/
e. Change my life	1 ▯	0 ▯	63/
f. Other	1 ▯	0 ▯	64/
→ What? ______________________			65/

23. When did you enlist?

1 ▯ While I was in high school → Continue. 66/
2 ▯ Right after I left high school → Continue.
3 ▯ A while after I left high school → Go to Question 26 on Page 11.

24. What were your three most important reasons for enlisting while you were still in high school or right after you left high school? You can use the responses in Questions 19-22 for your answers.

Most important reason: ______________________________ 67-69/

Second most important reason: ______________________________ 70-72/

Third most important reason: ______________________________ 73-75/

CARD 02

Page 10

CARD 03 6-7/ 1-5/

25. When you were in high school, which military services did you consider? Check all that apply.

Service		Reason	Code
1☐ Army	Why?	2☐ Benefits	8-9/
		3☐ Job Assignment	10/
		4☐ Training	11/
		5☐ Image	12/
		6☐ Family Tradition	13/
		7☐ Other: ________________	14-15/
8☐ Navy	Why?	9☐ Benefits	16-17/
		10☐ Job Assignment	18/
		11☐ Training	19/
		12☐ Image	20/
		13☐ Family Tradition	21/
		14☐ Other: ________________	22-23/
15☐ Air Force	Why?	16☐ Benefits	24-25/
		17☐ Job Assignment	26/
		18☐ Training	27/
		19☐ Image	28/
		20☐ Family Tradition	29/
		21☐ Other: ________________	30-31/
22☐ Marine Corps	Why?	23☐ Benefits	32-33/
		24☐ Job Assignment	34/
		25☐ Training	35/
		26☐ Image	36/
		27☐ Family Tradition	37/
		28☐ Other: ________________	38-39/
29☐ Coast Guard	Why?	30☐ Benefits	40-41/
		31☐ Job Assignment	42/
		32☐ Training	43/
		33☐ Image	44/
		34☐ Family Tradition	45/
		35☐ Other: ________________	46-47/

CARD 03

Page 11

26. Why did you ultimately choose the Army? Check all that apply.

[1]☐ The Army was my first choice — 48/

The other military services …

- [2]☐ … would not take me — 49/
- [3]☐ … did not have the jobs I wanted — 50/
- [4]☐ … did not have the benefits I wanted — 51/
- [5]☐ … offered a smaller bonus — 52/

[6]☐ Other: ______________________________ 53-54/

INSTRUCTION

If you enlisted while you were still in high school or right after you left high school, go to Question 43 on Page 18.

If you waited a while after high school before enlisting, answer Question 27 below and continue.

Section C. The Decision to Enlist Now

27. Why did you wait a while after high school before enlisting? Mark yes or no for each item.

	Yes	No	
a. I got a job	1 ☐	0 ☐	55/
b. I went to college/vocational school	1 ☐	0 ☐	56/
c. I took some time off	1 ☐	0 ☐	57/
d. I didn't qualify	1 ☐	0 ☐	58/
e. Someone didn't want me to enlist	1 ☐	0 ☐	59/
→ Who? ______________________			60/
f. I wasn't interested in joining the Army	1 ☐	0 ☐	61/
g. The Army didn't have the job/options I wanted	1 ☐	0 ☐	62/
h. I was concerned about war/combat	1 ☐	0 ☐	63/
i. Other	1 ☐	0 ☐	64/
→ What? ______________________			65/

CARD 03

Page 12

CARD 04 6-7/ 1-5/

28. After not enlisting in high school or right after you left high school, did any of the following opportunities for experience interest you when you decided to enlist later? Check yes or no for each one.

	Yes	No	
a. Skills	1 ☐	0 ☐	8/
b. Travel	1 ☐	0 ☐	9/
c. Adventure	1 ☐	0 ☐	10/
d. Work experience	1 ☐	0 ☐	11/

29. After not enlisting in high school or right after you left high school, did any of the following people encourage you to enlist later? Check yes or no for each one.

	Yes	No	
a. Parents	1 ☐	0 ☐	12/
b. Spouse	1 ☐	0 ☐	13/
c. Other family	1 ☐	0 ☐	14/
d. Boyfriend/Girlfriend/Fiancé(e)	1 ☐	0 ☐	15/
e. Friends	1 ☐	0 ☐	16/
f. Recruiter	1 ☐	0 ☐	17/

30. After not enlisting in high school or right after you left high school, did any of the following benefits interest you when you decided to enlist later? Check yes or no for each one.

	Yes	No	
a. Money for education	1 ☐	0 ☐	18/
b. Bonus	1 ☐	0 ☐	19/
c. Benefits (like health care)	1 ☐	0 ☐	20/
d. Pay	1 ☐	0 ☐	21/

31. When you were in high school or right after you left high school, did people discourage you from enlisting?

1☐ Yes → Continue. 22/

0☐ No → Go to Question 33 on Page 13.

CARD 04

Page 13

32. When you decided to enlist, did the people who previously discouraged you from enlisting … Check yes or no for each one.

	Yes	No	
a. … change their mind?	1 ☐	0 ☐	23/
b. … have opinions that matter less to you?	1 ☐	0 ☐	24/
c. … no longer have an impact on your life?	1 ☐	0 ☐	25/

33. After not enlisting in high school or right after you left high school, did any of the following other reasons lead you to enlist later? Check yes or no for each one.

	Yes	No	
a. Desire to get away	1 ☐	0 ☐	26/
b. No jobs at home	1 ☐	0 ☐	27/
c. Only dead end jobs at home	1 ☐	0 ☐	28/
d. School didn't work out	1 ☐	0 ☐	29/
e. Patriotism/Serve my country	1 ☐	0 ☐	30/
f. Change my life	1 ☐	0 ☐	31/
g. Completed college	1 ☐	0 ☐	32/
h. Start my working life/career	1 ☐	0 ☐	33/
i. Other	1 ☐	0 ☐	34/
→ What? ____________________			35/

34. What were your three most important reasons for enlisting? You can use the responses in Questions 28-33 for your answers.

Most important reason: ______________________________ *36-38/*

Second most important reason: ______________________________ *39-41/*

Third most important reason: ______________________________ *42-44/*

CARD 04

Page 14

35. Think back to the events that led up to your current enlistment. Who initiated your first talk with an Army recruiter? Check one.

1☐ I contacted the Army recruiter first 45-46/
2☐ The Army recruiter contacted me first
3☐ I was with a friend/relative who was meeting with the recruiter
4☐ Other: ____________________

36. Think back to the events that led up to your current enlistment. Did you have any of the following types of contact with the recruiter? Check yes or no for each one.

	Yes	No	
a. Talked at school	1 ☐	0 ☐	*47/*
b. Talked at an Army-sponsored event	1 ☐	0 ☐	*48/*
c. Talked during a chance encounter in public	1 ☐	0 ☐	*49/*
d. Talked at the recruiting station	1 ☐	0 ☐	*50/*
e. Talked by phone	1 ☐	0 ☐	*51/*
f. Talked in a chat room	1 ☐	0 ☐	*52/*
g. A friend/relative introduced us	1 ☐	0 ☐	*53/*
h. Other	1 ☐	0 ☐	*54/*

→ What? ____________________ *55/*

37. How much time was there from your first serious contact with a recruiter to the time you signed your contract? If less than one month, enter 0.

Number of months: ☐☐ *56-57/*

CARD 04

Page 15

38. In the process leading up to your current enlistment, did you learn about the military from any of the following people? Check yes or no for each one.

	Yes	No	
a. Relative	1 ☐	0 ☐	*58/*
b. Friend	1 ☐	0 ☐	*59/*
c. Teacher or guidance counselor at school	1 ☐	0 ☐	*60/*
d. Coach	1 ☐	0 ☐	*61/*
e. Clergy.	1 ☐	0 ☐	*62/*
f. Judge, youth counselor, or someone working in law enforcement	1 ☐	0 ☐	*63/*
g. Someone else	1 ☐	0 ☐	*64/*
↳ Who? ______________			*65/*

39. In the process leading up to your current enlistment, did you learn about the military from any of the following Internet sources? Check yes or no for each one.

	Yes	No	
a. Banner or pop-up ad on a website	1 ☐	0 ☐	*66/*
b. Military website	1 ☐	0 ☐	*67/*
c. Other website	1 ☐	0 ☐	*68/*
d. Blog	1 ☐	0 ☐	*69/*
e. Online military games.	1 ☐	0 ☐	*70/*
f. Email from someone other than a recruiter	1 ☐	0 ☐	*71/*
g. Chat room	1 ☐	0 ☐	*72/*

CARD 04

Page 16

40. In the process leading up to your current enlistment, did you learn about the military from any of the following media sources? Check yes or no for each one.

	Yes	No	
a. TV ad	1 ☐	0 ☐	73/
b. Radio ad	1 ☐	0 ☐	74/
c. Magazine ad	1 ☐	0 ☐	75/
d. Newspaper display ad or help-wanted ad	1 ☐	0 ☐	76/
e. Movies.	1 ☐	0 ☐	77/
f. Books	1 ☐	0 ☐	78/
g. News or feature story on TV	1 ☐	0 ☐	79/

41. In the process leading up to your current enlistment, did you learn about the military in any of the following other ways? Check yes or no for each one.

	Yes	No	
a. Letter or postcard in the mail	1 ☐	0 ☐	80/
b. Information posted at school	1 ☐	0 ☐	81/
c. A military-sponsored event	1 ☐	0 ☐	82/
d. At a job fair or career fair	1 ☐	0 ☐	83/
e. Other	1 ☐	0 ☐	84/
→ What?________________________			85/

CARD 04

Page 17

CARD 05 6-7/ 1-5/

42. In the process leading up to your current enlistment, which of the other military services did you consider? Check all that apply.

[1]☐ I did not consider any other military service.			8/
[2]☐ Navy → Why?	[3]☐	Benefits	9-10/
	[4]☐	Job Assignment	11/
	[5]☐	Training	12/
	[6]☐	Image	13/
	[7]☐	Family Tradition	14/
	[8]☐	Other: ____________	15-16/
[9]☐ Air Force → Why?	[10]☐	Benefits	17-18/
	[11]☐	Job Assignment	19/
	[12]☐	Training	20/
	[13]☐	Image	21/
	[14]☐	Family Tradition	22/
	[15]☐	Other: ____________	23-24/
[16]☐ Marine Corps → Why?	[17]☐	Benefits	25-26/
	[18]☐	Job Assignment	27/
	[19]☐	Training	28/
	[20]☐	Image	29/
	[21]☐	Family Tradition	30/
	[22]☐	Other: ____________	31-32/
[23]☐ Coast Guard → Why?	[24]☐	Benefits	33-34/
	[25]☐	Job Assignment	35/
	[26]☐	Training	36/
	[27]☐	Image	37/
	[28]☐	Family Tradition	38/
	[29]☐	Other: ____________	39-40/

CARD 05

Page 18

INSTRUCTION

All recruits continue by answering the questions below.

43. When did you sign your current enlistment contract? Month: ☐☐ Year: ☐☐☐☐ *41-46/*

44. In what state did you sign your current enlistment contract? State: ________________ *47-48/*

45. What was your ZIP code when you signed your current enlistment contract? ☐☐☐☐☐ *49-53/*

46. Did you require a waiver to join the Army? Check all that apply.

   1☐ No *54/*
   2☐ Yes, a conduct waiver (including moral or criminal). *55/*
      → How old were you when the offense occurred? ☐☐ years old *56-57/*
   3☐ Yes, a medical waiver *58/*
   4☐ Yes, a dependent waiver *59/*
   5☐ Yes, a drug abuse waiver *60/*
   6☐ Yes, another kind of waiver *61/*

47. With which component did you enlist? Check one.

   1☐ Active Army *62/*
   2☐ Army National Guard
   3☐ Army Reserve

48. Did any of the following benefits prompt you to choose that component? Check yes or no for each one.

	Yes	No	
a. Enlistment bonus	1 ☐	0 ☐	*63/*
b. Education benefits	1 ☐	0 ☐	*64/*
c. Other benefits	1 ☐	0 ☐	*65/*
d. Pay	1 ☐	0 ☐	*66/*

CARD 05

Page 19 (CARD 06) 6-7/ 1-5/

49. Did any of the following aspects of your military career prompt you to choose that component? Check yes or no for each one.

I wanted ...	Yes	No	
a. A higher entry pay grade	1 ☐	0 ☐	8/
b. To be a full-time soldier	1 ☐	0 ☐	9/
c. To be a part-time soldier	1 ☐	0 ☐	10/
d. Better long-term career opportunities	1 ☐	0 ☐	11/

50. Did any of the following location issues prompt you to choose that component? Check yes or no for each one.

I preferred ...	Yes	No	
a. Training near home	1 ☐	0 ☐	12/
b. Service near home	1 ☐	0 ☐	13/
c. Service away from home	1 ☐	0 ☐	14/

51. Did any of the following aspects of military assignment prompt you to choose that component? Check yes or no for each one.

I would be ...	Yes	No	
a. More likely to deploy	1 ☐	0 ☐	15/
b. Less likely to deploy	1 ☐	0 ☐	16/

52. Did any of the following other reasons prompt you to choose that component? Check yes or no for each one.

I wanted ...	Yes	No	
a. Skills	1 ☐	0 ☐	17/
b. Travel	1 ☐	0 ☐	18/
c. Adventure	1 ☐	0 ☐	19/
d. Work experience	1 ☐	0 ☐	20/
e. Other	1 ☐	0 ☐	21/
↳ What? ______________________			22/

CARD 06

Page 20

53. What do you think are the chances that you will deploy to a combat zone shortly after you complete your advanced individual training (AIT)? Check one.

23-24/

- 0 ☐ 0 in 10 (no chance)
- 1 ☐ 1 in 10 (almost no chance)
- 2 ☐ 2 in 10 (very unlikely)
- 3 ☐ 3 in 10 (unlikely)
- 4 ☐ 4 in 10 (less than even chance)
- 5 ☐ 5 in 10 (even chance)
- 6 ☐ 6 in 10 (better than even chance)
- 7 ☐ 7 in 10 (likely)
- 8 ☐ 8 in 10 (very likely)
- 9 ☐ 9 in 10 (almost certain)
- 10 ☐ 10 in 10 (certain)

54. Over the next 4 years, what do you think are the chances that you will deploy on more than one tour to a combat zone? Check one.

25-26/

- 0 ☐ 0 in 10 (no chance)
- 1 ☐ 1 in 10 (almost no chance)
- 2 ☐ 2 in 10 (very unlikely)
- 3 ☐ 3 in 10 (unlikely)
- 4 ☐ 4 in 10 (less than even chance)
- 5 ☐ 5 in 10 (even chance)
- 6 ☐ 6 in 10 (better than even chance)
- 7 ☐ 7 in 10 (likely)
- 8 ☐ 8 in 10 (very likely)
- 9 ☐ 9 in 10 (almost certain)
- 10 ☐ 10 in 10 (certain)

INSTRUCTION

If you enlisted in the Active Army, continue to Question 55 on Page 21.

If you enlisted in the Army Reserves or Army National Guard, go to Question 58 on Page 22.

CARD 06

Page 21

FOR RECRUITS WHO ENLISTED IN THE ACTIVE ARMY ONLY

55. Suppose you had enlisted in the Army Reserves or Army National Guard instead of the Active Army. What do you think your chances of deploying to a combat zone shortly after you completed your advanced individual training (AIT) would have been? Check one.

   0 ☐ 0 in 10 (no chance) *27-28/*
   1 ☐ 1 in 10 (almost no chance)
   2 ☐ 2 in 10 (very unlikely)
   3 ☐ 3 in 10 (unlikely)
   4 ☐ 4 in 10 (less than even chance)
   5 ☐ 5 in 10 (even chance)
   6 ☐ 6 in 10 (better than even chance)
   7 ☐ 7 in 10 (likely)
   8 ☐ 8 in 10 (very likely)
   9 ☐ 9 in 10 (almost certain)
   10 ☐ 10 in 10 (certain)

56. Suppose you had enlisted in the Army Reserves or Army National Guard instead of the Active Army. Over the next 4 years, what do you think your chances of deploying on more than one tour to a combat zone would have been? Check one.

   0 ☐ 0 in 10 (no chance) *29-30/*
   1 ☐ 1 in 10 (almost no chance)
   2 ☐ 2 in 10 (very unlikely)
   3 ☐ 3 in 10 (unlikely)
   4 ☐ 4 in 10 (less than even chance)
   5 ☐ 5 in 10 (even chance)
   6 ☐ 6 in 10 (better than even chance)
   7 ☐ 7 in 10 (likely)
   8 ☐ 8 in 10 (very likely)
   9 ☐ 9 in 10 (almost certain)
   10 ☐ 10 in 10 (certain)

57. How important was the chance that you would be deployed to a combat zone in your decision to enlist in the Active Army instead of the Army Reserves or Army National Guard? Check one.

   1 ☐ Not important *31/*
   2 ☐ Not very important
   3 ☐ Somewhat important
   4 ☐ Very important
   5 ☐ Extremely important

STOP. Go to Page 23.

CARD 06

Page 22

FOR RECRUITS WHO ENLISTED IN THE ARMY RESERVES OR ARMY NATIONAL GUARD ONLY

58. Suppose you had enlisted in the Active Army instead of the Army Reserves or Army National Guard. What do you think your chances of deploying to a combat zone shortly after you completed your advanced individual training (AIT) would have been? Check one.

32-33/

- 0 ☐ 0 in 10 (no chance)
- 1 ☐ 1 in 10 (almost no chance)
- 2 ☐ 2 in 10 (very unlikely)
- 3 ☐ 3 in 10 (unlikely)
- 4 ☐ 4 in 10 (less than even chance)
- 5 ☐ 5 in 10 (even chance)
- 6 ☐ 6 in 10 (better than even chance)
- 7 ☐ 7 in 10 (likely)
- 8 ☐ 8 in 10 (very likely)
- 9 ☐ 9 in 10 (almost certain)
- 10 ☐ 10 in 10 (certain)

59. Suppose you had enlisted in the Active Army instead of the Army Reserves or Army National Guard. Over the next 4 years, what do you think your chances of deploying on more than one tour to a combat zone would have been? Check one.

34-35/

- 0 ☐ 0 in 10 (no chance)
- 1 ☐ 1 in 10 (almost no chance)
- 2 ☐ 2 in 10 (very unlikely)
- 3 ☐ 3 in 10 (unlikely)
- 4 ☐ 4 in 10 (less than even chance)
- 5 ☐ 5 in 10 (even chance)
- 6 ☐ 6 in 10 (better than even chance)
- 7 ☐ 7 in 10 (likely)
- 8 ☐ 8 in 10 (very likely)
- 9 ☐ 9 in 10 (almost certain)
- 10 ☐ 10 in 10 (certain)

60. How important was the chance that you would be deployed to a combat zone in your decision to enlist in the Army Reserves or Army National Guard instead of the Active Army? Check one.

36/

- 1 ☐ Not important
- 2 ☐ Not very important
- 3 ☐ Somewhat important
- 4 ☐ Very important
- 5 ☐ Extremely important

CARD 06

Page 23

INSTRUCTION

All recruits continue by answering the questions below.

61. What is your grade? Check one.

[1] ☐ E-1 *37/*
[2] ☐ E-2
[3] ☐ E-3
[4] ☐ E-4

STOP. Wait for instructions.

62. What is your Military Occupational Specialty (MOS)? MOS code: ☐☐☐ *38-40/*

CARD 06

Page 24

Section D. Event History Calendar

To understand what you did before you enlisted, please fill in this calendar.

1999 or earlier	***2000***	***2001***	***2002***	***2003***	***2004***	***2005***	***2006***	***2007***	***2008***

Page 25

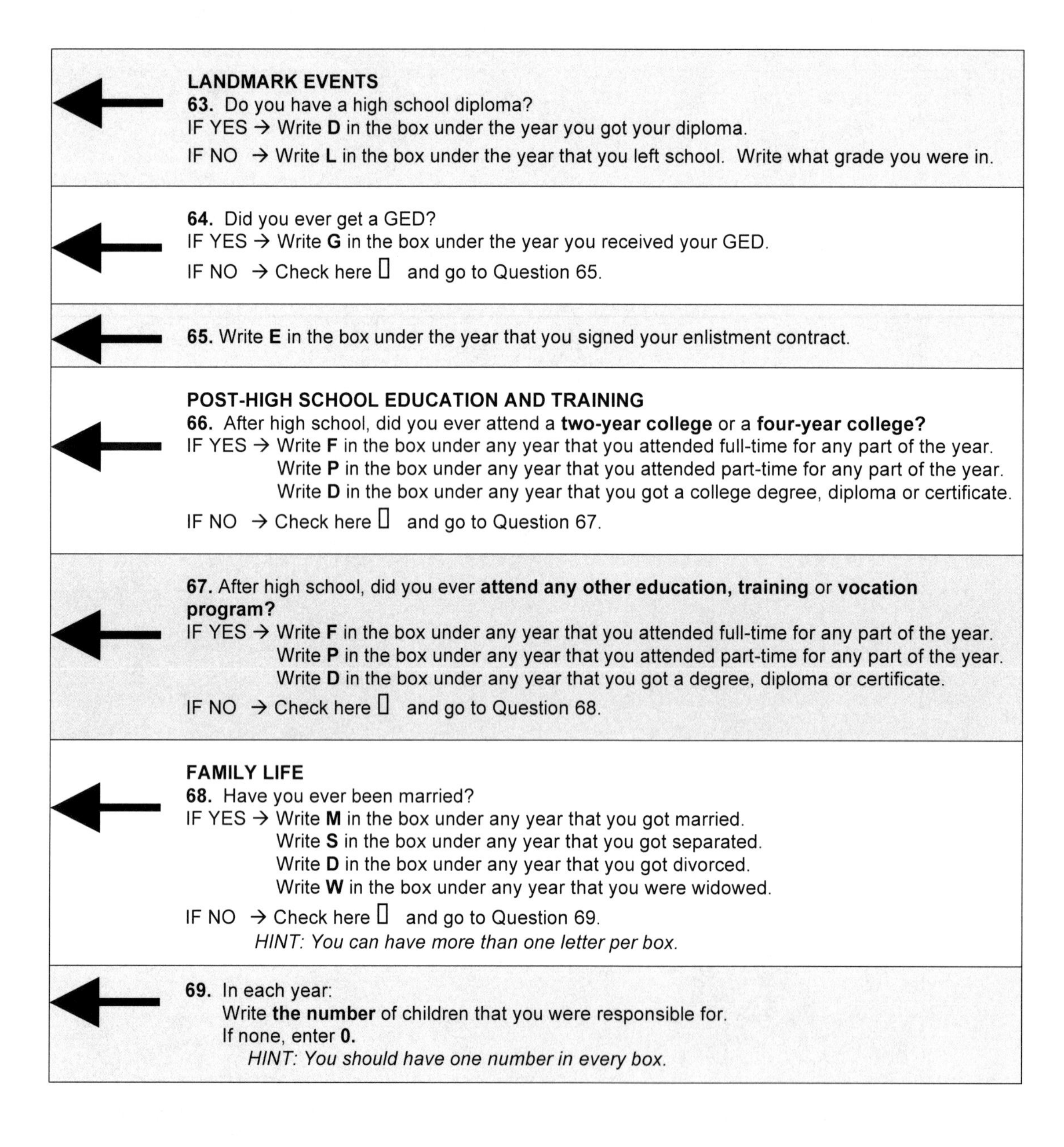

LANDMARK EVENTS
63. Do you have a high school diploma?
IF YES → Write **D** in the box under the year you got your diploma.
IF NO → Write **L** in the box under the year that you left school. Write what grade you were in.

64. Did you ever get a GED?
IF YES → Write **G** in the box under the year you received your GED.
IF NO → Check here ▯ and go to Question 65.

65. Write **E** in the box under the year that you signed your enlistment contract.

POST-HIGH SCHOOL EDUCATION AND TRAINING
66. After high school, did you ever attend a **two-year college** or a **four-year college?**
IF YES → Write **F** in the box under any year that you attended full-time for any part of the year.
Write **P** in the box under any year that you attended part-time for any part of the year.
Write **D** in the box under any year that you got a college degree, diploma or certificate.
IF NO → Check here ▯ and go to Question 67.

67. After high school, did you ever **attend any other education, training** or **vocation program?**
IF YES → Write **F** in the box under any year that you attended full-time for any part of the year.
Write **P** in the box under any year that you attended part-time for any part of the year.
Write **D** in the box under any year that you got a degree, diploma or certificate.
IF NO → Check here ▯ and go to Question 68.

FAMILY LIFE
68. Have you ever been married?
IF YES → Write **M** in the box under any year that you got married.
Write **S** in the box under any year that you got separated.
Write **D** in the box under any year that you got divorced.
Write **W** in the box under any year that you were widowed.
IF NO → Check here ▯ and go to Question 69.
HINT: You can have more than one letter per box.

69. In each year:
Write **the number** of children that you were responsible for.
If none, enter **0.**
HINT: You should have one number in every box.

Page 26

1999	2000	2001	2002	2003	2004	2005	2006	2007	2008

Page 27

EMPLOYMENT HISTORY

	70. Number of jobs
	71. State, territory or country
	72. Location (urban, suburban, rural)
JOB 1	**73.** Name of company
	74. Title/main job or task
	75. Hours worked per week
	76. Months worked at this job during this year
	77. Pay (per hour, day, week, etc.)
JOB 2	**78.** Name of company
	79. Title/main job or task
	80. Hours worked per week
	81. Months worked at this job during this year
	82. Pay (per hour, day, week, etc.)
	83. Number of months employed at a paying job
	84. Number of months not employed and looking for work
	85. Number of months not employed, not looking for work, but in school
	86. Number of months not employed, not looking for work, not in school

Thank you for completing the RAND Recruiting Survey.

If you have any comments, please provide them here: *41/*

CARD 06

Bibliography

Antel, John J., James Hosek, and Christine E. Peterson, *Military Enlistment and Attrition: An Analysis of Decision Reversal*, Santa Monica, Calif.: RAND Corporation, R-3510-FMP, 1987. As of February 19, 2013: http://www.rand.org/pubs/reports/R3510

Asch, Beth J., Paul Heaton, James R. Hosek, Francisco Martorell, Curtis Simon, and John T. Warner, *Cash Incentives and Military Enlistment, Attrition, and Reenlistment*, Santa Monica, Calif.: RAND Corporation, MG-950-OSD, 2010.
http://www.rand.org/pubs/monographs/MG950.html

Asch, Beth J., Paul Heaton, and Bogan Savych, *Recruiting Minorities: What Explains Recent Trends In The Army And Navy?* Santa Monica, Calif.: RAND Corporation, MG-861-OSD, 2009.
http://www.rand.org/pubs/monographs/MG861.html

Asch, Beth J., M. Rebecca Kilburn, and Jacob Alex Klerman, *Attracting College-Bound Youth Into the Military: Toward the Development of New Recruiting Policy Options*, Santa Monica, Calif.: RAND Corporation, MR-984-OSD, 1999.
http://www.rand.org/pubs/monograph_reports/MR984.html

Belli, R. F., "Computerized Event History Calendar Methods: Facilitating Autobiographical Recall," Paper Presented at American Statistical Association, Indianapolis, Ind.: August, 2000.

Bureau of Labor Statistics, "Longitudinal Surveys: The NLSY97," website, March 1, 2013. As of December 16, 2013:
http://www.bls.gov/nls/nlsy97.htm

DeGroot, Morris H., *Optimal Statistical Decisions*, New York: John Wiley and Sons, 1970.

Dertouzos, James N., *Recruiter Incentives and Enlistment Supply*, Santa Monica, Calif.: RAND Corporation, R-3065-MIL, 1985.
http://www.rand.org/pubs/reports/R3065.html

———, *Educational Benefits Versus Enlistment Bonuses: A Comparison of Recruiting Options*, Santa Monica, Calif.: RAND Corporation, MR-302-OSD, 1994.
http://www.rand.org/pubs/monograph_reports/MR302.html

———, *The Cost-Effectiveness of Military Advertising: Evidence From 2002–2004*, Santa Monica, Calif.: RAND Corporation, DB-565-OSD, 2009.
http://www.rand.org/pubs/documented_briefings/DB565.html

Diaconis, Persi, and Donald Ylvisaker, "Conjugate Priors for Exponential Families," *Annals of Statistics*, Vol. 7, 1979, pp. 269–281.

Fechter, Alan E. "Impact of Pay and Draft Policy on Army Enlistment Behavior," in *Studies Prepared for the President's Commission on an All-Volunteer Armed Force*, Washington, D.C.: U.S. Government Printing Office, 1970, pp. ii-3-1 to ii-3-59.

Freeman, Thornton D., "The Life History Calendar: A Technique For Collecting Retrospective Data," *Sociological Methodology*, Vol. 18, 1988, pp. 37–68.

Kilburn, M. Rebecca, and Beth J. Asch, *Recruiting Youth in the College Market: Current Practices and Future Policy Options*, Santa Monica, Calif.: RAND Corporation, MR-1093-OSD, 2003.
http://www.rand.org/pubs/monograph_reports/MR1093.html

Maddala, G. S., *Limited-Dependent and Qualitative Variables in Econometrics*, New York: Cambridge University Press, 1983.

McFadden, David, "Econometric Analysis of Qualitative Response Models," in Zvi and Michael Intrilligator, eds., *Handbook of Econometrics*, Vol. 2, Griliches, Amsterdam: Elsevier, 1983.

Milgrom, Paul R., "Good News and Bad News: Representation Theorems and Applications," *The Bell Journal of Economics*, Vol. 12, 1981, pp. 380–391.

National Longitudinal Program, "National Longitudinal Surveys," U.S. Bureau of Labor Statistics, U.S. Department of Labor, 2011. As of January 19, 2011:
http://www.bls.gov/nls/

Simon, Curtis J., Sebastian Negrusa, and John T. Warner, "Educational Benefits and Military Service: An Analysis of Enlistment, Reenlistment, and Veterans' Benefit Usage: 1991–2005," *Economic Inquiry*, Vol. 48, No. 4, October 2010, pp. 1008–1031.

Warner, John, and Curtis Simon., "Uncertainty about Job Match Quality and Youth Turnover: Evidence from U.S. Military Attrition," Working Paper, Clemson University, 2005.

Warner, John, Curtis Simon, and Deborah Payne, *Enlistment Supply in the 1990s: A Study of the Navy College Fund And Other Enlistment Incentive Programs*, Washington, D.C.: Defense Manpower Data Center, 2000-015, 2001.